MULTISENSOR SURVEILLANCE SYSTEMS
The Fusion Perspective

Related Recent Titles

Video-Based Surveillance Systems: Computer Vision and Distributed Processing
Paolo Remagnino, Graeme Jones, Nikos Paragios, Carlo Regazzoni
ISBN 0-7923-7632-3, 2002
http://www.wkap.nl/prod/b/0-7923-7632-3

Multimedia Video-Based Surveillance Systems: Requirements, Issues and Solutions
Gian Luca Foresti, Petri Mähönen, Carlo S. Regazzoni
ISBN 0-7923-7927-6, 2000
http://www.wkap.nl/prod/b/0-7923-7927-6

Biometrics: Personal Identification in Networked Society
Anil Jain, Ruud Bolle, Sharath Pankanti
ISBN 0-7923-8345-1, 1999
http://www.wkap.nl/prod/b/0-7923-8345-1

Advanced Video-Based Surveillance Systems
Carlo S. Regazzoni, Gianni Fabri, Gianni Vernazza
ISBN 0-7923-8392-3, 1999
http://www.wkap.nl/prod/b/0-7923-8392-3

MULTISENSOR SURVEILLANCE SYSTEMS
The Fusion Perspective

edited by

Gian Luca Foresti
University of Udine, Italy

Carlo S. Regazzoni
University of Genova, Italy

Pramod K. Varshney
Syracuse University, NY, USA

KLUWER ACADEMIC PUBLISHERS
Boston / Dordrecht / New York / London

Distributors for North, Central and South America:
Kluwer Academic Publishers
101 Philip Drive
Assinippi Park
Norwell, Massachusetts 02061 USA
Telephone (781) 871-6600
Fax (781) 871-6528
E-Mail: < kluwer@wkap.com>

Distributors for all other countries:
Kluwer Academic Publishers Group
Post Office Box 322
3300 AH Dordrecht, THE NETHERLANDS
Telephone 31 78 6576 000
Fax 31 78 6576 254
E-Mail: < services@wkap.nl>

 Electronic Services < http://www.wkap.nl>

Library of Congress Cataloging-in-Publication Data

Multisensor Surveillance Systems: The Fusion Perspective
Gian Luca Foresti, Carlo S. Regazzoni, Pramod K. Varshney (Eds.)
ISBN 1-4020-7492-1

Copyright © 2003 by Kluwer Academic Publishers

All rights reserved. No part of this work may be reproduced, stored in a retrieval system, or transmitted in any form or by any means, electronic, mechanical, photocopying, microfilming, recording, or otherwise, without prior written permission from the Publisher, with the exception of any material supplied specifically for the purpose of being entered and executed on a computer system, for exclusive use by the purchaser of the work.

Permission for books published in Europe: permissions@wkap.nl
Permissions for books published in the United States of America: permissions@wkap.com

Printed on acid-free paper.

Printed in the United States of America

CONTENTS

PREFACE ix

ACKNOWLEDGEMENTS xiii

I   MULTISENSOR FUSION IN SURVEILLANCE SYSTEMS 3
P.K. Varshney

1 A DISTRIBUTED SENSOR NETWORK FOR VIDEO
   SURVEILLANCE OF OUTDOORS...7
   G.L. Foresti and L. Snidaro

2 DISTRIBUTED METADATA EXTRACTION STRATEGIES IN A
   MULTIRESOLUTION DUAL CAMERA SYSTEM........................29
   L. Marcenaro, L.Marchesotti, and C.Regazzoni

3 AUTOMATIC TARGET ACQUISITION AND TRACKING WITH
   COOPERATIVE FIXED AND PTZ VIDEO CAMERAS.................43
   B. Abidi, A. Koschan, S. Kang, M. Mitckes, and M. Abidi

4 LEARNING THE FUSION OF VIDEO DATA STREAMS:
   AUTOMATIC CALIBRATION AND REGISTRATION OF
   SURVEILLANCE CAMERAS 61
   J. Renno, P. Remagnino, and G.A. Jones

5   IMAGE FUSION USING THE EXPECTATION-MAXIMIZATION ALGORITHM AND A GAUSSIAN MIXTURE MODEL 81

R. S. Blum and J. Yang

6   VIDEO-BASED SURVEILLANCE FOR CHEM-BIO PROTECTION OF BUILDINGS 97

I. Pavlidis, C. Stathopoulos, and T. Faltesek

II   DETECTION, TRACKING AND RECOGNITION 115

G.L. Foresti

7   SECOND GENERATION PREFILTERING FOR VIDEO COMPRESSION AND ANALYSIS IN MULTISENSOR SURVEILLANCE SYSTEMS 119

F. Ziliani and J. Reichel

8   ACQUIRING MULTI-VIEW VIDEO WITH AN ACTIVE CAMERA SYSTEM 135

R.T. Collins, O. Amidi, and T. Kanade

Contents

9   COMPUTATIONAL FRAMEWORK FOR SIMULTANEOUS REAL-TIME HIGH-LEVEL VIDEO REPRESENTATION – EXTRACTION OF MOVING OBJECTS AND RELATED EVENTS 149

A. Amer

10   MULTICAMERA SURVEILLANCE: OBJECT-BASED SUMMARIZATION APPROACH 183

F. Porikli

11   DETECTING DANGEROUS BEHAVIORS OF MOBILE OBJECTS IN PARKING AREAS 199

G.L. Foresti, G. Giacinto, and F. Roli

III   BIOMETRICS IN SURVEILLANCE SYSTEMS 215

C.S. Regazzoni

12   BIOMETRIC FEATURE EXTRACTION IN A MULTI-CAMERA SURVEILLANCE SYSTEM 219

S.L. Dockstader and A. M. Tekalp

13   FUSION OF FACE RECOGNITION ALGORITHMS FOR VIDEO-BASED SURVEILLANCE SYSTEMS 235

G.L. Marcialis and F. Roli

14 INFORMATION THEORY BASED FACE TRACKING 251

E. Loutas, C. Nikou, and I. Pitas

15 OPTIMUM FUSION RULES FOR MULTIMODAL BIOMETRIC SYSTEMS 265

L. Osadciw, P. Varshney, and K. Veeramacheneni

INDEX **287**

Preface

Monitoring of public and private sites is increasingly becoming a very important and critical issue, especially after the recent flurry of terrorist attacks including the one on the Word Trade Center in September 2001. It is, therefore, imperative that effective multisensor surveillance systems be developed to protect the society from similar attacks in the future. The new generation of surveillance systems to be developed have a specific requirement: they must be able to automatically identify criminal and terrorist activity without sacrificing individual privacy to the extent possible. Privacy laws concerning monitoring and surveillance systems vary from country to country but, in general, they try to protect the privacy of their citizens.

Monitoring and visual surveillance has numerous other applications. It can be employed to help invalids or handicapped and to monitor the activities of elderly people. It can be used to monitor large events such as sporting events, as well. Nowadays, monitoring is employed in several different contexts including transport applications, such as monitoring of railway stations and airports, dangerous environments like nuclear facilities or traffic flows on roads and bridges. The latest generation of surveillance systems mainly rely on hybrid analog-digital, or completely digital video communications and processing methods and take advantage of the greater flexibility offered by video processing algorithms that are capable of focusing a human operator's attention on a set of interesting situations.

Proposed solutions range from standard surveillance systems that are able to detect and track people moving in the observed scene to advanced surveillance systems based on multiple sensors/cameras (optical, infrared, thermal, radar, etc.) that are able to understand complex human behavior, to automatically detect and recognize their faces, and to discover their identity by means of specific biometric features.

Advanced surveillance systems use multiple sensors and, consequently, data fusion strategies, to increase both their field of view and their capability to detect interesting events. Multisensor surveillance systems can take advantage of either the same type of information acquired from different spatial locations or information acquired by sensors of the same type or different types on the same monitored area. Appropriate processing techniques and new sensors providing real-time information related to different scene characteristics can help both to enlarge the size of monitored environments and to improve performance in terms of activity detection over the areas monitored by the sensors.

Research work on multisensor real-time video processing techniques for robust video transmission, color-image processing, event-based attention focusing, model-based sequence understanding in surveillance applications is expected to provide important results that can be transitioned to practical systems. This has become feasible due to the availability of high computational power at acceptable costs.

This book combines well-established work with work in progress of international researchers from academia and industrial organizations. They have endeavored to solve theoretical problems related to the monitoring of a complex site with multiple sensors, and applied their techniques to real problems in different ensemble of environments for a number of different applications. This book aims at addressing several current issues in Multisensor Surveillance by providing a selected set of contributions by researchers from some of the leading laboratories intensively working in the surveillance field in different parts of the world.

The present book is the fourth volume in this series of books on video surveillance. The first book was entitled *Advanced Video-based Surveillance Systems* and was edited by C.S. Regazzoni, F. Fabris and G. Vernazza (1998). The second one was entitled *Multimedia Video-based Surveillance Systems: Requirements, Issue and Solution* and was edited by G.L. Foresti, P. Mahonen and C.S. Regazzoni (2000). The last published book in the series was entitled *Video-based Surveillance Systems: Computer Vision and Distributed Processing* and was edited by P. Remagnino, G. Jones, N. Paragios and C.S. Regazzoni (2002).

The current book is based on the important research results presented at two Special Sessions organized in 2002 by the book editors at two important International Conferences. The first Special Session entitled *Information Fusion Techniques for Surveillance and Security Applications* was organized by P.K. Varshney and G.L. Foresti at the International Conference on Information Fusion, July 7-11, Annapolis, Maryland, USA, while the second

Preface

one entitled *Multisensor Surveillance Systems* was organized by C.S. Regazzoni and P. K. Varshney at the IEEE International Conference on Image Processing, September 23-25, Rochester, New York, USA. The main objective of the Special Sessions on which the current book is based was to directly integrate end user needs with advanced research techniques. In this sense, the chapters of this book are expected to be of interest to a wide audience, ranging from end-users interested in surveillance systems for their specific purposes to academic and industrial researchers more oriented towards the advancement of their research and the development of cutting edge technology.

The book is structured into three sections: each of them is organized and introduced by one of the co-editors.

Section one covers the latest research on Multisensor Fusion issues in order to provide surveillance systems with augmented perception capabilities that can be used to detect a larger set of events.

Section two is focused on new techniques that are able to perform scene analysis as well as identify and track objects of interest along the field of view of multiple cameras.

Section three presents recent results in the field of biometrics with particular emphasis on new techniques that are able to perform face tracking and recognition.

Gian Luca Foresti

Carlo S. Regazzoni

Pramod K. Varshney

Acknowledgements

The editors wish to thank the authors for their high quality contributions describing their most recent research work and for their active and timely cooperation.

Special thanks go to Dr. Lauro Snidaro for his invaluable work of reformatting and finalizing the book. Moreover, we wish to thank a number of graduate students, research fellows and collaborators from the Artificial Vision and Real-Time Systems Lab at the Department of Computer Science of the University of Udine, particularly Dr. Chistian Micheloni and Claudio Piciarelli. We also express our appreciation to Dr. Alex Greene and Mrs. Melissa Sullivan from Kluwer Academic Publishers for their precious editorial support.

This work was partially supported by the Italian Ministry of University and Scientific Research within the framework of the project *"Distributed systems for multisensor recognition with augmented perception for ambient security and customization"* (2002-2004).

I

MULTISENSOR FUSION IN SURVEILLANCE SYSTEMS

MULTISENSOR FUSION IN SURVEILLANCE SYSTEMS

Pramod K. Varshney
Department of Electrical Engineering and Computer Science, Syracuse University, NY, USA

Multiple sensors are increasingly being used in visual surveillance systems. The use of multiple sensors enables us to augment the capabilities of single sensors in many aspects. The system can monitor a larger area, it can detect and reason about a larger set of events, and via the choice of appropriate sensors it can operate under a variety of difficult situations such as poor illumination and adverse weather conditions. This chapter focuses on current research activity on multisensor fusion for visual surveillance. It consists of six contributions, the first three describe ongoing research on the development of three specific visual surveillance systems while the next two present specific algorithms that can be employed in multisensor surveillance systems. The last contribution employs visual surveillance to attain a specific building security goal.

The first contribution by Foresti and Snidaro presents a distributed sensor network for video surveillance of outdoor environments. Their system consists of heterogeneous sensors to take advantage of the special properties of the sensors that complement each other. In particular, an optical and an infrared sensor are employed. Their performance complements each other under different lighting conditions and adverse weather conditions. They have described the architecture of their system and discussed the algorithms employed at each processing level. An example that provides results of each processing step on real data is provided. This contribution serves as an excellent introduction to video surveillance systems that employ heterogeneous sensors and fusion.

In the next contribution, Marcenaro, et al. describe their system for outdoor surveillance. This system employs a fixed camera and an active pan-tilt-zoom camera that work in a cooperative mode. The fixed camera captures low-resolution images of the entire scene. These are used to detect and locate moving objects. This information is used to determine the pan-tilt movement of the mobile camera to focus attention on the desired object at a higher zoom level. This distributed system is intelligent and achieves excellent performance in terms of probability of correct detection, false alarms and missed detections, and target location. The system operates at 2 frames/second. Research is underway to enhance the capabilities of the system by optimizing different algorithms, employing additional sensors and by incorporating additional intelligence.

In the third contribution, Abidi, et al. also present a video tracking and location system consisting of a fixed camera and a network of pan-tilt-zoom (PTZ) cameras. The main application of this system is to detect the motion of a subject going the wrong way. This is applicable to a variety of security applications including at airports and limited access buildings. The goal is to develop a digital, networked and fully automated system with cooperating cameras. The fixed camera is employed for breach detection. Once a breach is detected, the target is handed over to a PTZ camera for tracking and location processing. Preliminary results are reported and further development of the system is underway.

The next two contributions deal with the development of specific algorithms that can be used by multisensor visual surveillance systems. Renno et al. focus on automatic calibration and registration algorithms based on visual data acquired by multiple cameras. This suite of algorithms is such that the algorithms require minimal human intervention and are self-adjusting. The camera calibration approach consists of two stages. In the first stage the image-plane to local-ground-plane transformation of each camera is obtained. In the second stage, a clustering technique is used to recover the transformation between these ground planes. This algorithm has been shown to perform quite well by applying it to real data sets.

The next contribution of this chapter by Blum and Yang presents an image fusion approach based on a Gaussian mixture distortion model and employs the expectation-maximization (EM) algorithm. Unlike most other image fusion approaches, this approach is based on a rigorous application of estimation theory. An iterative EM-based algorithm is developed and is applied to the concealed weapon detection problem and the autonomous landing guidance application. Results show the superior performance of the image fusion algorithm.

In the final contribution of this chapter, Pavlidis et al. present a building protection system to protect against the specific threat of chemical and biological attacks on the air-intakes of the buildings. A layered security architecture is proposed where a multi-camera video surveillance system forms one of the layers. It is installed near building air-intakes and is employed to detect and monitor suspicious behavior. This information can be used to issue early warnings for evacuation and to initiate control and mitigation actions. Preliminary results on system performance are encouraging and enhancements in the human activity recognition module are currently being investigated.

Chapter 1

A DISTRIBUTED SENSOR NETWORK FOR VIDEO SURVEILLANCE OF OUTDOORS

G.L. Foresti and L. Snidaro
Department of Mathematics and Computer Science, University of Udine, Via delle Scienze 208, 33100 Udine, Italy.

Key words: DSN, Video surveillance

1. INTRODUCTION

New generation surveillance systems [1], [2], [3] require to manage large amounts of visual data (optical, infrared, etc.). Recently, the development of sensor technology and computer networks has contributed to increase the interest in Distributed Sensor Networks (DSNs) for real-time information fusion [4], [5], [6].

DSNs are basically systems composed of a set of sensors, a set of processing elements (PEs), and a communication network interconnecting the PEs. In this paper, a DSN architecture for a video surveillance system, where sensors may be either physical (i.e., to generate information by translating observed physical quantities into electrical signals) or virtual (i.e., to produce new information from existing one), is proposed. PEs fuse data acquired by different physical and/or virtual sensors, or pre-processed by other PEs at lower levels in order to reduce the degree of uncertainty naturally associated with the acquired information, and to produce an interpretation of the environment observed. The dimensionality of information is reduced as only significant information is propagated through the network.

The proposed DSN architecture integrates optical and infrared sensors to support 24 hours per day a real-time visual-based surveillance system for outdoor environments. IR and optical sensors are at the first level of the proposed architecture. Video signals of each physical sensor are first processed to extract moving image regions, called *blobs* [7], and features are computed for the target tracking, classification, and data fusion procedures. This integration allows to improve at higher levels the accuracy of object localization which is based on the ground plane hypothesis [7] and object recognition [8]. At the first level, specialized PEs track each detected blob on the image plane and transform 2D blob positions (in the sensor coordinates system) into 3D object positions (in the coordinates of the monitored environment's map). Each first level PE is committed to the surveillance of a sub-area of the monitored environment.

The trajectory of each blob, extracted by a given sensor, is first approximated with cubic splines [7]. Splines are used to save bandwidth as they can represent a trajectory with a very limited number of points. In this way only spline parameters are needed to be sent through the network thus avoiding to transmit the targets' positions at every time instant.

At the higher level of the architecture a trajectory fusion of the local object trajectories is performed to compute the trajectory of the objects with respect to the whole map of the monitored site. Information about object trajectories and blob features can be used to learn a neural network (e.g., a neural tree [8], [9]) to recognize suspicious events in the observed scene [10], [11].

A real example of the proposed DSN architecture employing infrared and optical sensors will be presented in the context of the video surveillance of an outdoor parking area.

2. DSN ARCHITECTURE

DSN architectures are a debated topic since early 80s [12]. A review of the recent advancements can be found in [13] and [21]. Although general discussions on the advantages and disadvantages of the various network topologies are vastly present in the literature, little work has been done on DSNs for video surveillance systems. These are characterized by sources generating great amounts of data at high frequency (typically 25 fps). Therefore, some of the parameters to be considered in choosing the most appropriate network structure are the following:

1. A distributed sensr network for video surveillance of outdoors 9

- data to be sent;
- network bandwidth;
- data fusion technique;
- efficiency;
- cost.

The overall system architecture is shown in Fig. 1-1. The network has a tree structure where the lowest level (leaves) is constituted by heterogeneous (optical, black&white, IR, etc.) multi-resolution sensors. Nodes represent processing elements.

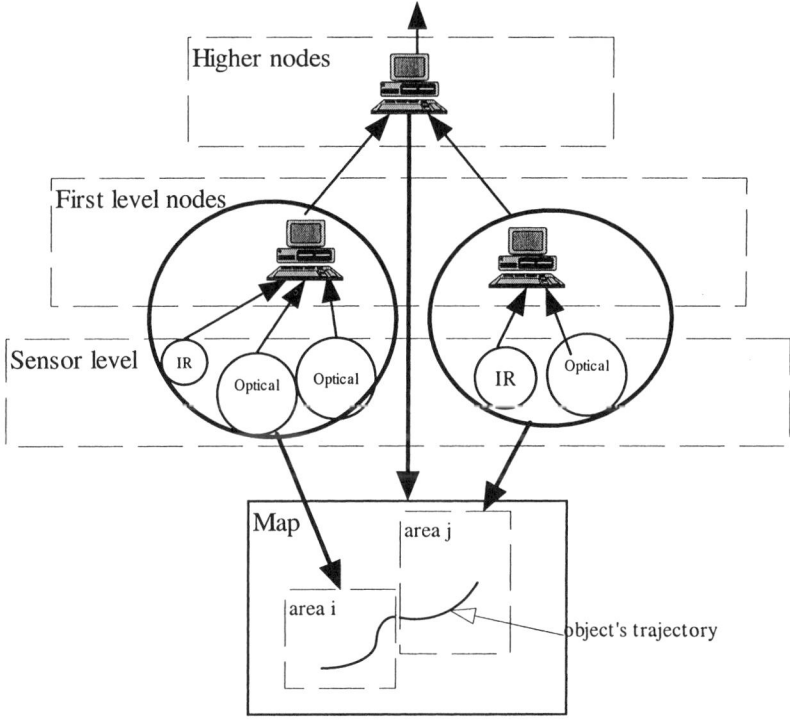

Figure 1-1. The proposed DSN architecture

Internal nodes (PEs) are subdivided into First Level nodes and Higher level nodes. The former are directly connected to the sensors, forming the so called *clusters* [13], and are in charge of the monitoring of sub areas of the environment under surveillance. The latter receive data from the connected First Level nodes for broader area coverage.

The tree topology has been chosen as it offers a hierarchical flow of data and it's easier to extend than the general anarchic committee (AC) structure [12] where no hierarchy is defined. Intrinsically more vulnerable then a fully interconnected scheme, the tree topology can be made more fault tolerant adding a certain degree of redundancy to the inner nodes (PEs). Table 1-1 summarizes some of the characteristics of the possible architectures along with their key strengths and weaknesses.

Table 1-1. Architecture comparison

	Centralized	Hierarchical	Distributed
Characteristics	Single fusion site Single track database	Hierarchical structure	No central fusion site No fixed structure
Advantages	Theoretically optimal Simple structure	Sensors naturally partitioned Simple fusion algorithm	Highly fault tolerant Processing load is distributed
Disadvantages	High processing load High bandwidth requirements Low reliability	suboptimal	Complicated fusion and communication algorithms non-optimal

Hierarchical in nature, the chosen topology allows an intuitive partitioning of the sensors according to the different sub-areas of the environment to be monitored. More sophisticated architectures such as the flat tree [14] or the deBruijin graph [15] could be more difficult and expensive to implement, and may even not be useful since with the proposed approach there is no need to exchange data between nodes of the same level. A fully distributed scheme is the most robust one but also the most complicated. Sophisticated fusion and communication algorithms have to be developed. This architecture is also non-optimal in nature. An in-depth comparison and discussion can be found in [21]. A detailed description of the chosen network structure is given in the following sections.

2.1 Sensors

Each cluster comprises one or more static cameras. Heterogeneous sensing technologies (optical, infrared, etc.) may be employed to assure the surveillance system's operational state during day and night and in presence of changing weather conditions. Active sensors can also be employed in a cooperative fashion as in [16], [20].

1. A distributed sensr network for video surveillance of outdoors

If smart sensors (with image processing capabilities) are available, video signals are locally pre-processed to perform the blob extraction procedure. This step pinpoints moving regions in the image through change detection algorithms well-known in the literature [1], [2], [17], [18], [7], [19]. A number of features can be extracted from the blobs (discussed later in Section 3.2).

Blobs and their features are sent through the network to the connected first level node. Note that the amount of information sent through the network at this level is relatively small (blobs are generally much smaller than the whole image) and can be handled efficiently by current network technology thus allowing the connection of many sensors to a single node.

Figure 1-2. Video signals obtained from a color and an IR camera.

The segmentation and feature extraction modules, being at the beginning of the processing chain, are therefore crucial to the overall performance of the system and care has to be taken in implementing them. In particular, outdoor environments are particularly tricky to deal with due to changing illumination and weather conditions, light spots, clutter, etc. For this reason multiple heterogeneous sensors and data fusion techniques are employed to guarantee the performance of the system in every condition. Figure 1-2 shows the video signals obtained from a color and an IR camera at night and in presence of fog.

2.2 First level nodes

First-level nodes perform object tracking within the field of view of the connected sensors (Figure 1-3). Each cluster can be considered committed to the surveillance of a particular area. Thus, each of them has a set of sensors monitoring the same area (the field of view is -partially- overlapping). Therefore, target tracking is performed taking advantage of the redundancy offered by the multiple readings coming from the connected sensors. When a target is in the field of view of two or more sensors, extracted data from each view can be associated and fused together since they are representing the same object. Obviously, more than one target can be present in the field of view of the sensors, and all have to be tracked. This problem is known in the literature as multi-sensor multi-target tracking and has been studied in detail [24], [25].

The tracking/fusion process involves the following functions:

- data alignment;
- data association;
- updating.

The first deals with the spatial registration of the data in a common reference system. For a video surveillance system this generally implies the conversion of the image plane coordinates into the 2D map coordinates of the area being monitored. The second is responsible for the partitioning of the measurements into sets that could have originated from the same targets. The third step leads to the updating of each target track with the measurements obtained from the sensors.

First level nodes can also implement recognition procedures to classify moving objects in order to improve the tracking performance [8], [16].

The trajectory analysis procedure maintains the positions assumed by objects on the 2D top view map of the area and calculates cubic splines to approximate their trajectory. Spline segments (along with targets' features used by the tracking algorithms) are then sent to the supervising higher level node. This approach considerably reduces bandwidth requirements as will be discussed in Section 3.5

1. A distributed sensr network for video surveillance of outdoors 13

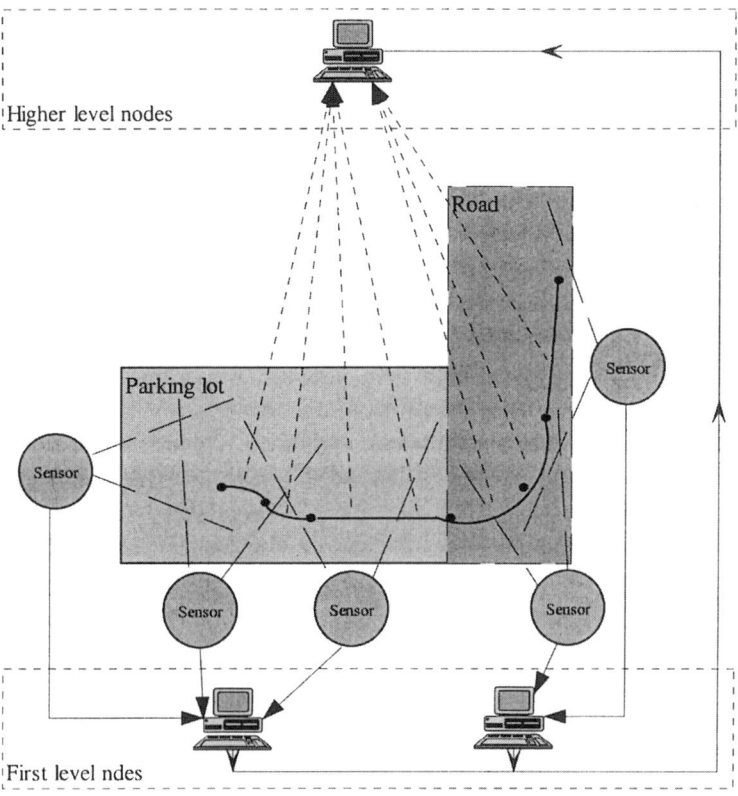

Figure 1-3. Two clusters monitoring different areas.

2.3 Higher level nodes

Targets' position, information (identifier, dimensions, class,...) and trajectory (cubic splines) are sent from first level nodes to the higher level nodes. These nodes are committed to a wider region that should include all the areas covered by the first level nodes attached to them. They are responsible for trajectory analysis throughout the controlled sub-areas, that is, they can examine a target's path as a whole (Figure 1-3). These nodes are therefore in charge for event detection and alarm triggering. This is the higher software layer that constitutes an automatic surveillance system. Since this paper is more focused on the lower levels of the system, the reader is referred to [10] and [11]. Data fusion can also be performed at this level for decision making based on hypothesis [22], [23].

Higher level nodes also constitute the interface with the operator as they implement a graphical user interface providing information on the position of the targets, warning level, events, etc. A good example of the functionalities of a modern user interface for a video surveillance system is given in [16].

No visual data is moved between first and high level nodes during the normal operation of the system. Video signals are brought to the operator's attention only when an alarm is triggered or a suspicious event has been detected.

3. PROCESSING

This section describes the processing steps that take place in each component of the proposed architecture. The flow-chart in Fig. 1-4 shows the flow of data through the nodes and the procedures involved; the configuration considered consists of two heterogeneous sensors (optical and infrared) connected to a first level node which in turn feeds a higher level node.

3.1 Blob extraction

This processing step occurs at sensor level. Motion detection and blob extraction is exploited following a layered background subtraction approach [16]. Change detection is performed using an algorithm for automatic threshold computation based on Euler numbers [19]. The background is updated using a Kalman filter [8]. Frame by frame subtraction is also applied in conjunction to improve detection results [16]. Morphological filters are also applied to improve the quality of the extracted blobs by removing spurious pixels due to noise, and by enhancing the regions' connectivity [16].

Following the scheme in Fig. 1-4, at time t, images **VI**(t) and **IR**(t) are produced respectively by the optical and infrared sensor. Each sensor applies the blob extraction procedure thus obtaining blobs $\boldsymbol{B}^{j}_{vi}(t)$ and $\boldsymbol{B}^{k}_{ir}(t)$ with $1 \leq j \leq n_{vi}$ and $1 \leq k \leq n_{ir}$ where n_{vi} and n_{ir} are the number of blobs extracted on the current frame respectively by the optical and the infrared sensor.

1. A distributed sensr network for video surveillance of outdoors

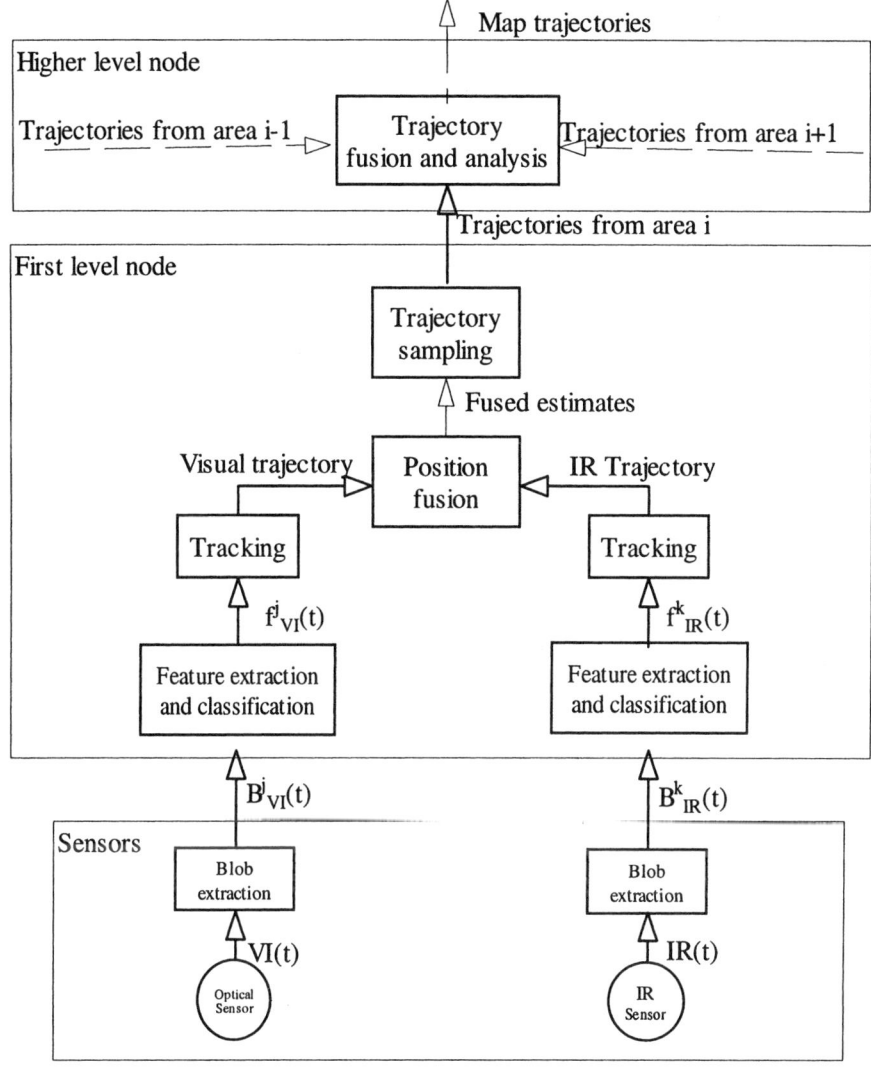

Figure 1-4. Processing steps

In Figure 1-5 are shown the blobs extracted from the video signals obtained from a color and an IR camera respectively.

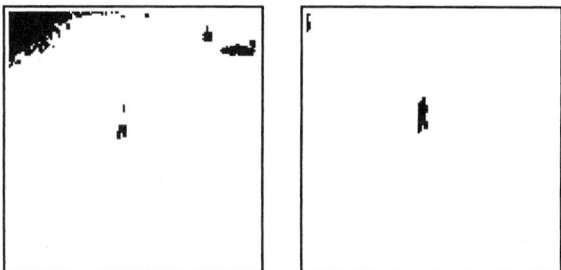

Figure 1-5. Blobs extracted from the images shown in Figure 1-2.

3.2 Feature extraction and classification

Blobs collected from sensors undertake further processing at first level nodes. In particular, some features are needed to describe the moving objects. Typical measures are area, perimeter, dispersedness ($perimeter^2 / area$), and centroid coordinates of the blob, along with measurements regarding the dimensions of the blob's bounding box (base, height, base/height ratio). Additional information may be acquired computing and analyzing the color histograms of the blobs [31], [32].

This information (in the form of the feature vector $f^i(t)$, where i spans the detected objects) may be used by the module deputed to object recognition [8], [16]. A classifier (e.g. a neural tree [9]) may be employed to identify the objects hence adding precious data to the feature vector characterizing the object. The more information is filled in the feature vector, the more robust will be the subsequent target tracking procedure.

3.3 Target tracking

In a video surveillance system, usually multiple objects exist in the scene. For example, in a parking lot during daytime, there could be many objects, such as people and vehicles, that move around.

The system needs to maintain tracks for all objects simultaneously. Hence, this is a typical multi-sensor multi-target tracking problem: measurements should be correctly assigned to their associated target tracks and a target's associated measurements from different sensors should be fused to obtain better estimation of the target state.

A first tracking process may occur locally to each image plane. For each sensor, the system can perform an association algorithm to match the current detected blobs with those extracted in the previous frame. A number of techniques is available, spanning from template matching, to features matching [16], to more sophisticated approaches [32].

When the position of the targets has to be estimated by fusing the observations of different sensors, map tracking is more than a choice. A common reference system is in fact needed to align the data extracted by each camera.

Generally, a 2D top view map of the monitored environment is taken as a common coordinates system [7], [8], but even the GPS may be employed to globally pinpoint the targets [16]. The former approach is obviously more straightforward to implement, as a well-known result from projective geometry states that the correspondence between an image pixel and a planar surface is given by a planar homography [29], [30]. The pixel usually chosen to represent a blob and be transformed into map coordinates is projection of the blob's centroid on the lower side of the bounding box [7], [8], [16].

Once the blob has been localized on the map, it must be associated with the known tracks at the previous frame or instantiated as a new one.

To deal with the multi-target data assignment problem, especially in the presence of persistent interference, there are many matching algorithms available in the literature: Nearest Neighbor (NN), Joint Probabilistic Data Association (JPDA), Multiple Hypothesis Tracking (MHT), and S-D assignment. The choice depends on the particular application; detailed descriptions and examples can be found in [24], [33], and [34].

The tracking procedure may be rather complex if all the information contained in the feature vector $\vec{f}(t)$ (Section 3.2) has to be considered in the matching phase. Moreover, this procedure is expected to be smart enough to not be fooled by noise or occlusions. Recent developments on the subject may found in [16], [35], [36], [37], and [38].

3.4 Position fusion

Data obtained from the different sensors (extracted features) can be combined together to yield a better estimate. A typical feature to be fused is the target's position on the map. The theoretically optimal fusion scheme involves the fusion of the positions of the target (according to the different sensors) obtained right out of the coordinate conversion function (Figure 1-5(a)). This is simple measurement fusion through Kalman filter [26]. Although optimal, this approach is difficult to employ in practice. In fact, measurements can be widely varying due to noise. In particular, they can be heavily affected by segmentation errors. Take for example the processed image on the left of Figure 1-2: the silhouette of the person is not completely extracted. If the centroid of the blob was to be transformed into the object's position on the map, the error would be consistent. This may occur every time the blob is erroneously segmented (due noise, lighting conditions, etc.) thus leading to an inaccurate target's localization and trajectory on the map. This problem may be alleviated running a Kalman filter for each track to obtain a filtered estimate of the target's position thus smoothing high variations due to segmentation errors (Figure 1-5(b)).

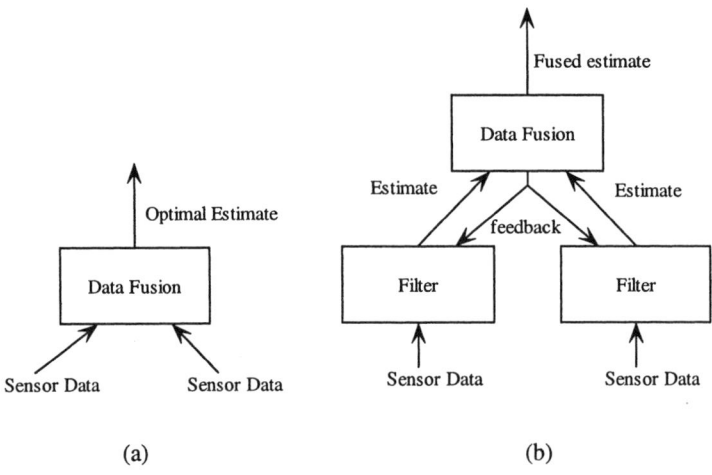

Figure 1-6. (a) Optimal fusion and (b) track fusion

When the latter approach is chosen, the fused estimate has to be based on the outputs of the track filters through track-to-track fusion techniques (the reader is referred to [27] and [28] for recent developments). The following equations for track fusion (without and with feedback) are based on the information form of the Kalman filter [21].

Without feedback:

$$P^{-1}(t|t) = P^{-1}(t|t-1) + \sum_{i=1}^{N}\left[P_i^{-1}(t|t) - P_i^{-1}(t|t-1)\right] \quad (1.1)$$

$$P^{-1}(t|t)\hat{x}(t|t) = P^{-1}(t|t-1)\hat{x}(t|t-1) + \\ + \sum_{i=1}^{N}\left[P_i^{-1}(t|t)\hat{x}_i(t|t) - P_i^{-1}(t|t-1)\hat{x}_i(t|t-1)\right] \quad (1.2)$$

1. A distributed sensr network for video surveillance of outdoors

With feedback:

$$P^{-1}(t|t) = \sum_{i=1}^{N} P_i^{-1}(t|t) - (N-1)P^{-1}(t|t-1) \qquad (1.3)$$

$$P^{-1}(t|t)\hat{x}(t|t) = P^{-1}(t|t-1)\hat{x}(t|t-1) + \\ + \sum_{i=1}^{N} \left[P_i^{-1}(t|t)\hat{x}_i(t|t) - P_i^{-1}(t|t-1)\hat{x}_i(t|t-1) \right] \qquad (1.4)$$

In the above equations, $\hat{x}_i(t|t)$, $\hat{x}_i(t|t-1)$, $P_i(t|t)$, and $P_i(t|t)$ denote the updated and predicted estimates and error covariances for the lower filters i, while the nonsubscripted quantities denote those of the fusion filter. Note that performing data fusion with feedback involves the decorrelation of the local estimates as they are affected by a common process noise and cannot be considered independent.

3.5 Trajectory sampling

This module receives fused estimates for each object's map position on in the area controlled by the PE and transmits their trajectories to a higher level node. To reduce bandwidth requirements, this module samples objects' positions every z time intervals and calculates the cubic spline passing trough the points $M(k)$ and $M(k+z)$. The points between $M(k)$ and $M(k+z)$ are given by:

$$M_k(t) = M_k + M'_k t + \left[\frac{3(M_{k+z} - M_k)}{t^2_{k+z}} - \frac{2M'_k}{t_{k+z}} - \frac{M'_{k+z}}{t_{k+z}} \right] t^2 + \\ + \left[\frac{2(M_k - M_{k+z})}{t^3_{k+z}} + \frac{M'_k}{t^2_{k+z}} + \frac{M'_{k+z}}{t^2_{k+z}} \right] t^3 \qquad (1.5)$$

where M_k and M_{k+z} are object positions at time t and $t+z$. M'_k and M'_{k+z} are the tangent vectors in M_k and M_{k+z}. t_k and t_{k+z} are the parameter values at the beginning and end of the curve segment. Fig. 1-7 shows a curve formed by two cubic splines linked together.

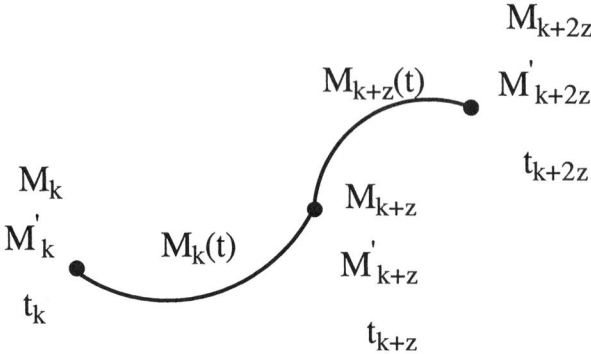

Figure 1-7. A curve formed by two spline segments.

Therefore, only the starting and ending points and tangent vectors are needed to define a spline segment. Each of them may span a predefined number of measurements (track points on the map) or, to improve bandwidth savings, a dynamically computed one. In particular, the sampling procedure may take into account whether the target is maneuvering or not and regulate the sampling frequency accordingly (the trajectory of a maneuvering target requires more spline segments to be described). The spline segments are then sent to the higher level nodes.

3.6 Trajectory fusion and analysis

The trajectory fusion occurs at higher level PE. These nodes are in charge of composing the trajectories of the targets by joining together the spline segments sent by the first level PE. The joining itself poses no problem when the segments are provided by a single first level PE, as every spline segment is received along with the target's ID so that the association is easily performed. When a target moves from a sub-area to another controlled by a different first level PE, an association algorithm (Section 3.3) is required to maintain the target's ID at the higher level PE.

The higher level nodes are the ideal processing elements for running trajectory analysis algorithms as they have a broader view of the behavior of the targets. Depending on the specific application, the system can detect and signal to the operator dangerous situations (e.g. a pedestrian walking towards a forbidden area, a vehicle going in a zigzag or in circles, etc.) [10], [11].

4. RESULTS

A color camera and a b/w camera with near infrared response have been employed for experiments. Image grabbing was performed at 256x256 pixels resolution. The cameras have been placed to monitor the same area of the front courtyard at Rizzi building of the University of Udine.

Figure 1-7 reports the trajectory of the person in Figure 1-2 according to the color (a) and IR (b) cameras. The trajectory computed by the system is plotted in yellow dots while the trajectory plotted in black denotes the ground truth path covered by the walking person. Figure 1-8 shows the results obtained through data fusion.

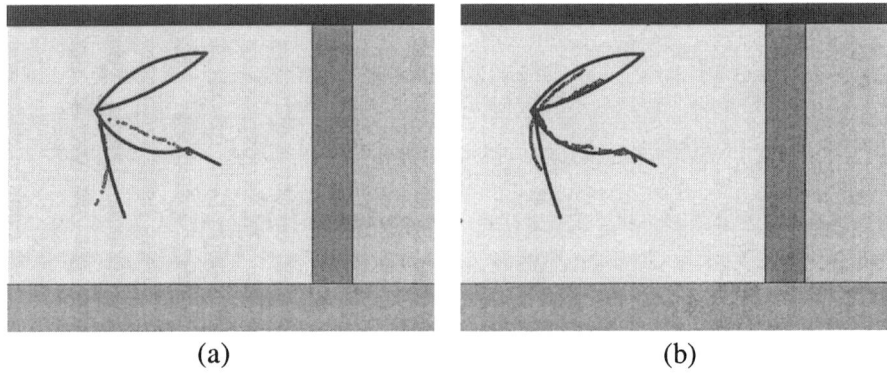

(a) (b)

Figure 1-8. Trajectory of the person in Figure 1-2 according to the (a) color and (b) IR cameras.

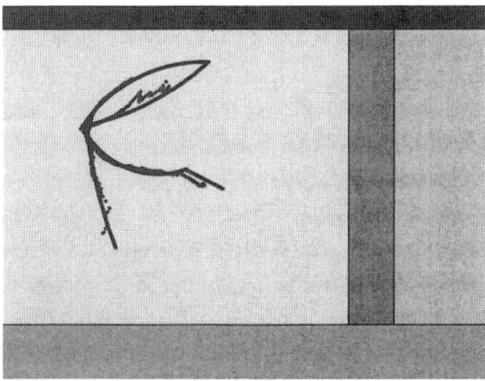

Figure 1-9. Trajectory of the person in Figure 1-2 obtained through data fusion.

The video sequence was acquired at night in presence of fog. The area was irradiated by an IR beacon to obtain infrared response from the b/w camera.

As can be seen comparing the two images of Figure 1-7, the results obtained with the color camera only are poor as the trajectory is discontinuous and affected by segmentation errors (Figure 1-5). Figure 1-7(b) shows the trajectory computed using the IR video signal only. As expected, the trajectory is more continuous and close to the ground truth. However, the target is still temporarily lost (top right) as it traverses a dense fog bank. The fusion process yields the results shown in Figure 1-8: the trajectory is similar to the IR one but takes into account also the positions obtained from the color camera. The fused estimates are even closer to the ground truth data.

Table 1-2. Mean and standard deviation (in pixels) of the distance between estimated and ground truth positions for the trajectories of Figure 1-7 and 1-8.

	Mean	Standard deviation
IR camera	8.66	4.42
Color camera	17.22	9.28
Data fusion	8.58	4.03

Table 1-2 reports the mean and standard deviation of the distance between estimated and ground truth positions. The IR camera is surely performing better than the color one and the data fusion procedure yields even better results (although very close to the IR camera, as expected).

In the following experiment the two cameras are employed in daylight conditions. Two persons are walking in the scene.

Figure 1-10 shows the source and processed frames for the b/w and color cameras respectively. In Fig. 1.12 the trajectory obtained by the fusion procedure is shown. It is more continuous than the ones produced by the single sensors (Figure 1.11), especially when compared with the B/W one. Moreover, the fused trajectory is more similar to the ground truth as confirmed by Tables 1-3 and 1-4. This is due to the reduction of calibration errors through data fusion.

Note that the color camera is not performing better than the b/w for each frame; as shown in Figure 1-10, blob "A" has been extracted better from the b/w video signal. This is indicative of the effectiveness of the heterogeneous sensor fusion approach.

During the tests the trajectories were performed by an average number of 7 splines segments. The average number of measurements composing each track was roughly 1000. So, only 28 points per track were sent to the higher level node instead of 1000.

1. A distributed sensr network for video surveillance of outdoors

Figure 1-10. Source and processed frames for the b/w and color cameras.

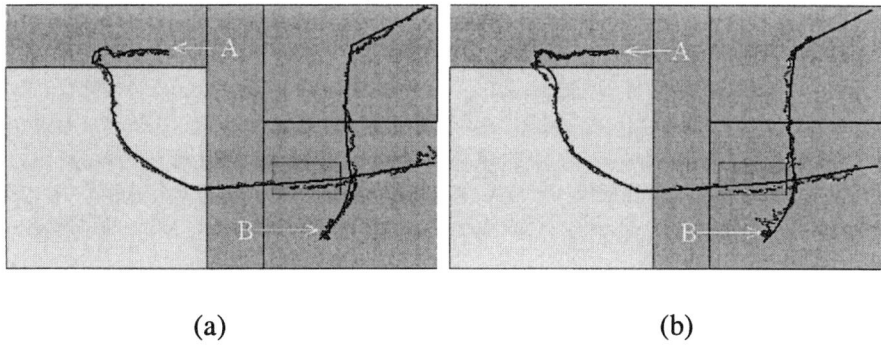

Figure 1-11. Trajectories of the two persons in Figure 1-10 according to the (a) color and (b) b/w cameras.

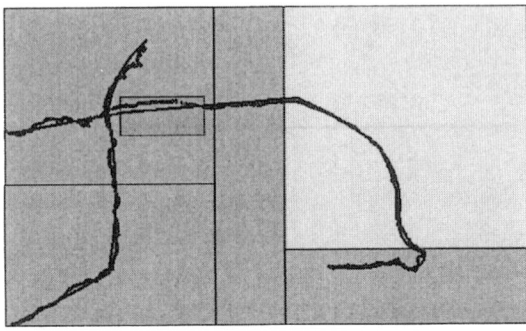

Figure 1-12. Trajectory of the person in Figure 1-10 obtained through data fusion.

Table 1-3. Mean and standard deviation (in pixels) of the distance between estimated and ground truth positions for the trajectories of blob "A" in Figures 1-10 and 1-11.

	Mean	Standard deviation
IR camera	10.64	4.75
Color camera	8.94	3.62
Data fusion	8.21	3.02

Table 1-4. Mean and standard deviation (in pixels) of the distance between estimated and ground truth positions for the trajectories of blob "B" in Figures 1-10 and 1-11.

	Mean	Standard deviation
IR camera	11.2	5.23
Color camera	9.83	4.67
Data fusion	8.61	3.79

5. CONCLUSIONS

We presented a distributed sensor network for video surveillance capable of day and night operation even in presence of adverse weather conditions. The system takes advantage from heterogeneous multi-resolution sensors by performing data fusion at various levels. Due to the redundant information, the system is able to track the targets more efficiently and accurately.

The proposed architecture also reduces networks requirements and is easily scalable and maintainable.

The system has been tested for the video surveillance of an outdoor area under different lighting and weather conditions.

REFERENCES

[1] C. Regazzoni, V. Ramesh, and G.L. Foresti, "Special issue on video communications, processing, and understanding for third generation surveillance systems", *Proceeding of the IEEE*, vol. 89, n. 10, 2001.

[2] R.T Collins, A.J. Lipton, H. Fujiyoshi, and T. Kanade, "Special Section on Video Surveillance", *IEEE Transactions of Pattern Analysis and Machine Intelligence*, Vol. 22, n. 8, August 2001.

[3] G.L. Foresti, P. Mahonen and C.S. Regazzoni, Multimedia Video-Based Surveillance Systems: from User Requirements to Research Solutions, Kluwer Academic Publishers, 2000.

[4] D.N. Jayasimha, S.S. Iyengar, and R.L. Kashyap, "Information integration and synchronization in distributed sensor netwoks", *IEEE Transactions on System, Man, and Cybernetics*, Vol. 21, n. 21, pp. 1032-1043, Sept/Oct 1991.

[5] A. Knoll and J. Meinkoehn, "Data fusion using large multi-agent networks: an analysis of network structure and performance", in Proceedings of the International Conference on Multisensor Fusion and Integration for Intelligent Systems (MFI), pp.113-120, Las Vegas, NV, Oct.2-5 1994.

[6] H Qi, S. Iyengar, and K. Chakrhbarty, "Multiresolution data integration using mobile agents in distributed sensor networks", *IEEE Transaction on Systems, Man, and Cybernetics – Part C: Applications and Reviews*, Vol. 31, No. 3, pp. 383-391, 2001.

[7] G.L.Foresti, "Real-time detection of multiple moving objects in complex image sequences", *International Journal of Imaging Systems and Technology*, Vol. 10, pp. 305-317, 1999.

[8] G.L. Foresti, "Object recognition and tracking for remote video surveillance", *IEEE Transaction on Circuits and Systems for Video Technology*, Vol. 9, No. 7, 1999, pp. 1045-1062.

[9] G.L. Foresti, C. Micheloni, and L. Snidaro, "Adaptive high order neural trees for pattern recognition", in Proceedings of the International Conference on Pattern Recognition (ICPR), Quebec City, Canada, August 2002.

[10] T. Wada and T. Matsuyama, "Multiobject behavior recognition by event driven selective attention method", *IEEE Transactions on Pattern Analysis and Machine Intelligence*, Vol.22, n. 8, pp. 873-887, August 2000.

[11] G. Medioni, I. Cohen, F. Brémond, S. Hongeng and R. Nevatia, "Event detection and analysis from video streams", *IEEE Transactions on Pattern Analysis and Machine Intelligence,* Vol. 23, n. 8, pp. 873-889, August 2001.

[12] R. Wesson, F.Hayes-Roth, J.W. Burge, C.Stasz, and C.A. Sunshine, "Network structures for distributed situation assessment", *IEEE Transactions on System, Man, and Cybernetics*, Vol. 11, n. 1, pp. 5-23, January 1981.

[13] H.Qi, S. S. Iyengar, K. Chakrabarty, "Distributed Snesor Networks – A review of recent research", *Journal of the Franklin Institute*, Vol. 338, pp. 729-750, March 2001.

[14] L. Prasad, S. S. Iyengar, R. L. Kashyap, and R. N. Madan, "Functional characterization of sensor integration in distributed sensor networks", *IEEE Transactions on System, Man, and Cybernetics*, Vol.21, n. 5, pp. 1082-1087, Sept./Oct. 1991

[15] S.S. Iyengar, D.N. Jayasimha, and D. Nadig, "A versatile architecture for the distributed sensor integration problem", *IEEE Transactions on Computers*, Vol. 43, n. 2, pp. 175-185, February 1994.

[16] R.T. Collins, A.J. Lipton, H. Fujiyoshi and T. Kanade, "A system for video surveillance and monitoring," Proceedings of the IEEE, Vol. 89, pp. 1456-1477, October 2001.

[17] K. Skiestad and R. Jain, "Illumination independent change detection for real world image sequences", *Computer Vision Graphics and Image Processing*, vol. 46, pp. 387-399, 1989.

[18] P.L. Rosin and T. Ellis, "Image difference threshold strategies and shadow detection", in *Proceedings of the 6th British Machine Vision Conference*. 1995, pp. 347-356, BMVA Press.

[19] L. Snidaro and G.L. Foresti, "Real-time thresholding with Euler numbers", *Pattern Recognition Letters*, Vol. 24, n. 9-10, pp. 1533-1544, June 2003.

[20] C. Micheloni, G.L. Foresti, and L. Snidaro, "A cooperative multi-camera system for video-surveillance of parking lots", in *Proceedings of the IDSS conference*, London, February 2003.

[21] M. E. Liggins II, C. Chong, I. Kadar, M. G. Alford, V. Vannicola, S. Thomopoulos, "Distributed fusion architectures and algorithms for target tracking", *Proceedings of the IEEE*, Vol. 85, n. 1, pp. 95-107, January 1997.

[22] Z. Chair and P. K. Varshney, "Optimal data fusion in multiple sensors detection systems", *IEEE Transactions on Aerospace and Electronic Systems*, Vol. AES-22, pp.98-101, January 1986.

[23] P.K. Varshney, "Distributed Detection and Data Fusion", Springer-Verlag, 1997

[24] Y. Bar-Shalom and X. Li, "Multitarget-Multisensor Tracking: Principles and Techniques", YBS Publishing, 1995.

[25] Y. Bar-Shalom and W. D. Blair (editors), "Multitarget multisensor tracking: applications and advances Volume III", Artech House, 2000.

[26] D. Willner, C.B. Chang, and K.P. Dunn, "Kalman filter algorithms for a multi-sensor system", Proceedings of the IEEE Conference on Decision and Control, pp. 570-574, 1978.

[27] K.C. Chang, R.K. Saha, and Y. Bar-Shalom, "On optimal track-to-track fusion", IEEE Transactions on Aerospace and Electronic Systems, Vol. AES-33, n. 4, pp. 1271-1275, October 1997.

[28] K.C. Chang, Zhi Tian, and R.K. Saha, "Performance evaluation of track fusion with information filter", in Proceedings of the International Conference on Multisource-Multisensor Information Fusion, pp. 648-655, July 1998.

[29] R. Tsai, "A versatile camera calibration technique for high-accuracy 3D machine vision metrology using off-the-shelf TV cameras and lenses", *IEEE Journal of Robotics and Automation*, Vol. RA-3, n. 4, pp. 323-344, 1987.

[30] O.D. Faugeras, Q.-T. Luong, S.J. Maybank "Camera Self-Calibration: Theory and Experiments," *Proceedings of European Conference on Computer Vision*, pp. 321-334, 1992.

[31] Gary Bradski, "Computer Vision Face Tracking for Use in a Perceptual Interface", Intel Technology Journal, 2nd Quarter, 1998.

[32] Dorin Comanesciu, V. Ramesh and Peter Meer, "Real-time Tracking of Non-Rigid Objects using Mean Shift," IEEE Conference on Computer Vision and Pattern Recognition, Hilton Head, South Carolina, 2000.

[33] S.S. Balckman, "Multiple-target tracking with radar applications", Artech House, 1986

[34] A.B. Poore, "Multi-dimensional assignment formulation of data association problems arising from multi-target and multi-sensor tracking", *Computational Optimization and Applications*, Vol. 3, pp- 27-57, 1994.

[35] Alan Lipton, "Local Application of Optic Flow to Analyze Rigid vs Non-Rigid Motion," in ICCV Workshop on Frame-Rate Vision, Corfu, Greece, September 1999.

[36] I. Mikic, S. Santini, and R. Jain, "Tracking objects in 3d using multiple camera views", in *Proceedings of ACCV*, January 2000.

[37] S.L. Dockstader and A.M. Tekalp, "Multiple Camera Tracking of Interacting and Occluded Human Motion", *Proceedings of the IEEE*, Vol. 89, n. 10, pp. 1441-1455, 2001.

[38] A. Mittal and L. S. Davis, "M2tracker: A multi-view approach to segmenting and tracking people in a cluttered scene", *International Journal of Computer Vision*, vol. 51, no. 3, Feb/March 2003.

Chapter 2

DISTRIBUTED METADATA EXTRACTION STRATEGIES IN A MULTI-RESOLUTION DUAL CAMERA SYSTEM

L.Marcenaro, L.Marchesotti and C.S.Regazzoni
Department of Biophysical and Electronical Engineering - University of Genova

Key words: Automatic video-surveillance, Multi-camera system, Cooperative system

1. INTRODUCTION

Several advanced video-surveillance systems based on video processing and understanding techniques have been recently developed [1, 2]. The principal aim of such systems is to recognize and classify potentially dangerous situations and consequentially generate some kind of alarm to raise the attention of a human operator. The use of automatic scene understanding systems is becoming more and more frequent in modern society: in particular, video-surveillance systems can be used for transport monitoring [3, 4], urban and building security [5], tourism [6], and bank protection [7, 8], even if their use was originally restricted to a military related field [9, 10]. Fast improvements in computing capabilities, cheap sensors and advanced image processing algorithms can be considered as the enabling technologies for the development of real-time video surveillance and monitoring systems. In particular, aspects related to the distribution of intelligence in cooperative systems need to be considered for the development of third-generation surveillance systems. A multiple sensors setup can be useful for satisfying several requirements on the system functionalities: a system using several video sensors without overlapped fields of view can be useful when a large area needs to be guarded. In this case the understanding system should be able to integrate observations from different sensors on a spatio-temporal basis for extracting the correlation between the different points of view and generate a augmented tracking by following the tracked object as it passes from the view-point of one sensor to another. Industrial systems such as DETER [11] have been proposed for monitoring large open spaces, like parking lots, and report unusual moving patterns by pedestrian or vehicles. On the other hand, a system using multiple sensors with overlapped fields of view can be used when a complex scene has to be automatically monitored. If the guarded area presents many environmental occlusions or it is characterized by a high number of objects simultaneously present in the acquired images, the correlation between data acquired by using multiple cameras can be used in order to improve the correctness of the results of the scene understanding algorithms. Many data fusion techniques have been studied and developed [12] in order to integrate observed data in a common reference system. Data fusion techniques can be used in this case for improving the performances of scene understanding algorithms as demonstrated in [13]. Unfortunately, video-surveillance cameras are typically very expensive because good lenses and good response to poor illumination conditions are usually required to be able to work automatically in a wide range of real outdoor conditions. Beside this, one should consider that each camera needs to be connected to a frame grabber device and that typically a general purpose PC can hardly handle more than two frame grabbers if a high processing rate has to be guaranteed. The problem can be solved by using a mobile-head camera, e.g. a camera mounted on a pan/tilt unit that can be controlled by a standard PC. The remote PC can be often used to command optical parameters of the camera such as zoom, focus, shutter and iris aperture [14]. The adopted solution consists of two general purpose computer using a fixed and a mobile camera respectively. The system that processes images acquired through fixed camera is able to detect and track objects entering the guarded scene. Extracted information about moving objects are sent to the remote system that is equipped with a mobile camera. Pan-tilt-zoom parameters are estimated that are necessary to focus the attention of the system on a particular event in the scene. A well determined cooperation strategy is used in order to decide the most important event in the scene at a certain time instant. The paper is organized as follows: Section 2 describes the system architecture; in particular logical and physical architectures of the system are described. Section 3 focuses on the cooperation strategy that has been adopted for the considered system; Section 4 describes joint calibration procedures while Section 5 presents and analyze the achieved experimental results. Finally, conclusions are drawn in Section 6.

2. SYSTEM ARCHITECTURE

The proposed system can be decomposed in three basic layers each containing collaborative, distributed modules specifically devoted to well defined tasks; in the first layer the process of acquisition of data from sensors is performed, in the second the information is processed and then metadata related to moving objects (blobs) is extracted. In the last layer the state (position) of the sensors is updated in order to maximize the information content of metadata previously evaluated. The system architecture has been designed by keeping in consideration two main issues:
- analysis of the monitored scene with different levels of resolution (with the possibility of focusing the attention of the system on a particular region).
- monitor a wider area of interest.

To give an high-level overview of the system, it is described in terms of physical and logical architecture.

2.1 Physical architecture

The physical architecture is sketched in fig. 1; it is composed by two computational units (750Mhz PentiumIII based PC) both connected to CCD pan-tilt cameras. In particular an outdoor surveillance camera as been used as static sensor to monitor the entire scene, whereas the second camera is equipped with a mobile head unit and acts as active sensor that can be pan/tilt/zoom controlled through the host PC. Frame grabbers in the two PCs acquire images in PAL format (768x576) with 24bits color depth; images are actually processed at 2 frames per second. A client/server approach has been used to connect the two PCs and to enable the cooperation of the two sensors with standard TCP/IP communication channels. This solution permits to decentralize computational units and to locate sensors in the more appropriate sites without logistic constraints. Test took place making use of IEEE 802.11 wireless LAN interfaces to validate the possibility of having a remote sensor that does not need any wired link. In this sense the camera and the corresponding PC can be viewed as an "intelligent" sensor with the capability of broadcasting high-level metadata to a central processing unit that will be able to further process and integrate these data.

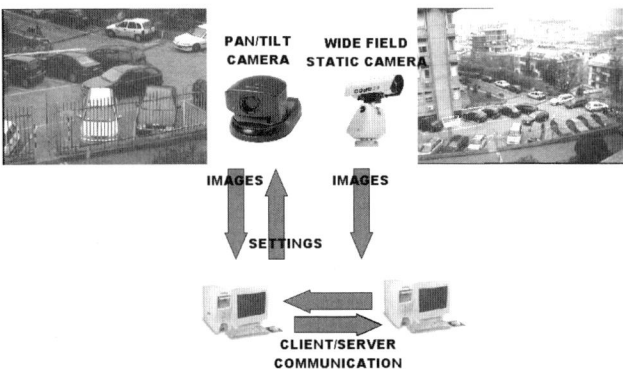

Figure 2-1. The Physical Architecture of the system.

2.2 Logical architecture

The logical architecture is made up by different processing tasks that can be grouped in representation, recognition and communication modules. The hierarchy of modules showed in Fig. 2-1 represents the logical configuration of the system using two cooperating PCs. The aim of this configuration is to have a system capable of monitoring a scene with a wide field of view, extracting salient information of the moving object and then focus the attention (acquiring frames with higher resolution) through a second mobile camera. The first part of the chain of modules on PC1 (connected to the static camera) reflects the scheme of a classical Video Surveillance systems, in which low-level representation modules operate at pixel level grabbing images from the camera ("Acquisition module"), evaluating difference images ("Change Detection module") based on a background image dynamically updated [15] and performing some morphological filtering in order to enhance image quality. The "Blob Coloring" module in PC1 processing chain follows the "Change Detection" module and acts as interface between low-level Image Processing modules and interpretation, high-level representation modules [16] providing a synthetic representation of region of amorphous pixels by using bounding boxes.

2. Distributed metadata extraction strategies in a multiresolution dual camera system

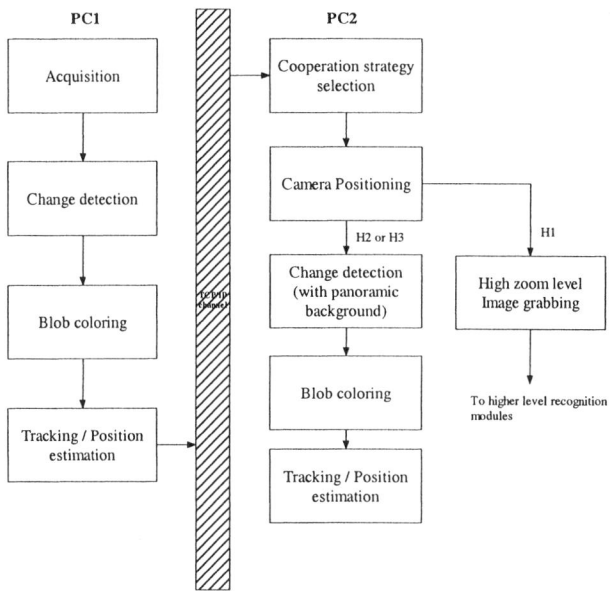

Figure 2-2. The logical architecture of the system.

Tracking module is able to identify each moving object in the scene on subsequent frames. The last module of PC1 give as output an estimation of the position of each blob in 2D image coordinates and 3D world coordinates. Positions of moving objects are sent to the second PC by using a TCP/IP communication channel: a Communication module has been implemented in order to activate a full duplex TCP/IP connection between the two processing units. In PC2 a chain of logical modules has been allocated in order to receive position data from PC1, to point the second camera on particular "targets" (blobs) and to successfully acquire a magnified view of moving objects following an optimized policy that will be described in section 2. The "Cooperation Strategy" module can be considered as a Recognition Module. It uses data extracted by representation modules a-priori event model (pointing strategy) in order to take a decision about a particular behavior of the system. Once PC2 gets position of a blob in terms of 3-D World Coordinates (the coordinates of the center of mass of the blob), the "Cooperation Strategy" module has to take a decision about the action to be performed by using mobile camera subsystem. Then pan and tilt angle to position the moving camera are evaluated by the "Camera Pointing" module. If an intermediate zoom level for the mobile sensor is chosen, the next step to be performed is related to the selection of a particular region of interest in the panoramic background off-line generated by the mobile camera [17] in order to detect moving blobs from the mobile camera. In this way the comprehension level of a scene can be increased in terms of blobs detection and trajectories estimation by using information acquired from zoomed point of view. If the "Cooperation Strategy" chooses that a high zoom-level has to be applied (virtual gate crossing), mobile head camera subsystem grabs the corresponding image of the area guarded at a higher zoom and passes the image to a central control unit that could be able for example to guess the identity of a person, by using face detection and recognition techniques.

3. COOPERATION STRATEGIES

This subsection will consider the "Cooperation Strategy" module of the proposed system in details. One of the main principal function of a automatic video-surveillance system is to detect and successfully monitor events of interest such as group of persons in a particular area, presence of a vehicle or the transit of people through "virtual gates". The proposed system approaches this issue with an innovative method to perform the analysis of such events. In particular, the system, once an event of interest is detected tries to acquire a more detailed representation of the scene by fine tuning the position and the zoom of the moving camera. The system can operate with two different modalities, automatically selected depending on the event detected:
- Acquisition and objects detection with intermediate zoom
- Acquisition with high zoom

In order to fully understand the functioning modalities of the proposed system, a state diagram is proposed in Fig. 2-3 where what we call "activation strategy" is represented.

The following notation will be used in the following:
- H0: no event (mobile head sensor subsystem is idle)
- H1: virtual gate trespassing
- H2: new object detected
- H3: old object successfully tracked by static camera

In order to get H3 additional conditions have to be verified: the temporal and spatial displacements of the tracked object with respect to the previously homologous object have to be Δt and $(\Delta x, \Delta y)$ respectively. H1 has the highest priority while other events (*H2, H3, H4*) have decreasing priorities.

Once the 3-D coordinates of a moving object are received by mobile sensor subsystem, it switches in "high zoom" mode and points the camera toward a particular zone of the scene where the virtual gate is located if the moving object is in the proximity of this particular area. Otherwise (*H3* and *H4*) the system goes in "intermediate" zoom modality and subsequent tests are carried out on event temporal and spatial information. If a new moving object is detected, pan-tilt unit is pointed toward that particular object, while if only old objects are detected in the scene, the mobile camera attention is focused onto the object only if at least Δt seconds are passed from the last event related to that object and the object displacement from its old position is $(\Delta x, \Delta y)$ at least.

This strategy is intended to minimize mobile head movements: pan/tilt and zoom movements are very expensive from a temporal point of view because it can take quite a long time in order to let the mobile sensor to change its pointing direction. Because of this it is not possible to continuously track moving objects by using the moving camera only. Instead the moving video sensor can be used when a higher detail is needed on a certain event in the guarded scene. The error probability of the system can be defined in case of "virtual gate trespassing" event as $P_{err} = p(H1 | H0, H2, H3)$: in this case an empty or not significant high zoom image is grabbed and sent to a higher level module for face detection and recognition. This action can be very expensive for the system functioning and has to be avoided if possible.

The probability of correct detection for a certain event Hx can be defined as $P_d^{Hx} = p(Hx | Hx \text{ is present})$ while the probability of false alarm for event Hx is defined as $P_{Fa}^{Hx} = p(Hx | Hx \text{ is not present})$.

4. VIDEO SENSORS CALIBRATION

In this subsection, a more detailed description of techniques used for joint video sensors calibration is given.

The static video sensor is calibrated by using standard Tsai algorithm. Then "Camera Pointing" module is able to associate pan/tilt angles to 3D position of objects estimated by using static camera subsystem. This is achieved by applying Tsai calibration algorithm to association (f,t) / ( ) for finding the matrix K able to get pan and tilt angles from 3D object coordinates.

At least 12 pairs (f,t) / ( ) need to be considered in order to precisely estimate "positioning calibration" matrix K.

A more complex technique is used in order to get world coordinates of objects detected by using the mobile sensor.

Mobile camera is calibrated by using the same set of points used for static sensor but the considered image coordinates are now referred to the global panoramic background of the guarded area. By using this strategy, when a object is detected at a intermediate zoom level by the mobile sensor, its image coordinate within that particular video shot are rescaled in the panoramic background reference system.

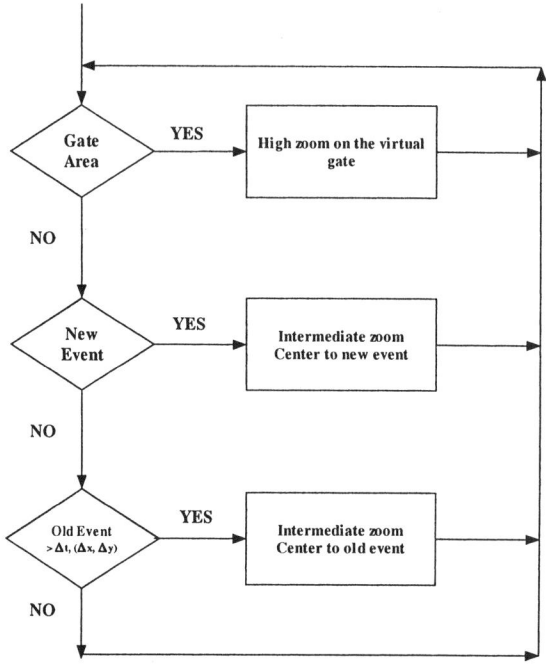

Figure 2-3. Cooperation strategy diagram.

By using the previously computed calibration matrix associating image coordinated of the panoramic background with 3D coordinates it is possible to estimate 3D objects position from the images at an intermediate zoom level. By using this strategy a more precise trajectory is computed because, for example, in many cases a group of objects detected as a single entity by the static camera are actually splitted by the mobile video sensor.

5. RESULTS

This section describes results achieved by the proposed system. Figure 2-4 and 2-5 show two different situations: an object entering the "virtual gate area" (2-4a) and the related high-level zoom video shot (2-4b) and a group of two people (2-5a) solved by the medium level video-shot (2-5b-c).

2. Distributed metadata extraction strategies in a multiresolution dual camera system

(a)   (b)

Figure 2-4. A pedestrian is trespassing a gate: (a) image acquired from the static camera; (b) image acquired from mobile camera

Figure 2-6 shows an example of the panoramic background that is generated by using the medium level zoom and that allows change detection by using mobile head camera.

Three different sequences have been considered for estimating the probabilities of correct detection and error of the proposed system. The probabilities of correct detection, false alarm and error have been estimated by analyzing the behavior of the system in different situations. The probability of correct detection can be defined as the probability of detecting an object when the object is actually present in the scene. The probability of false alarm takes into account the number of time when an object is detected even if it is not present, while error probability is related to a wrong activation of the mobile camera subsystem: this should evaluate the activation strategy goodness.

It has to be noticed that by analyzing a finite number of sequences and situations only a probability estimate can be carried out.

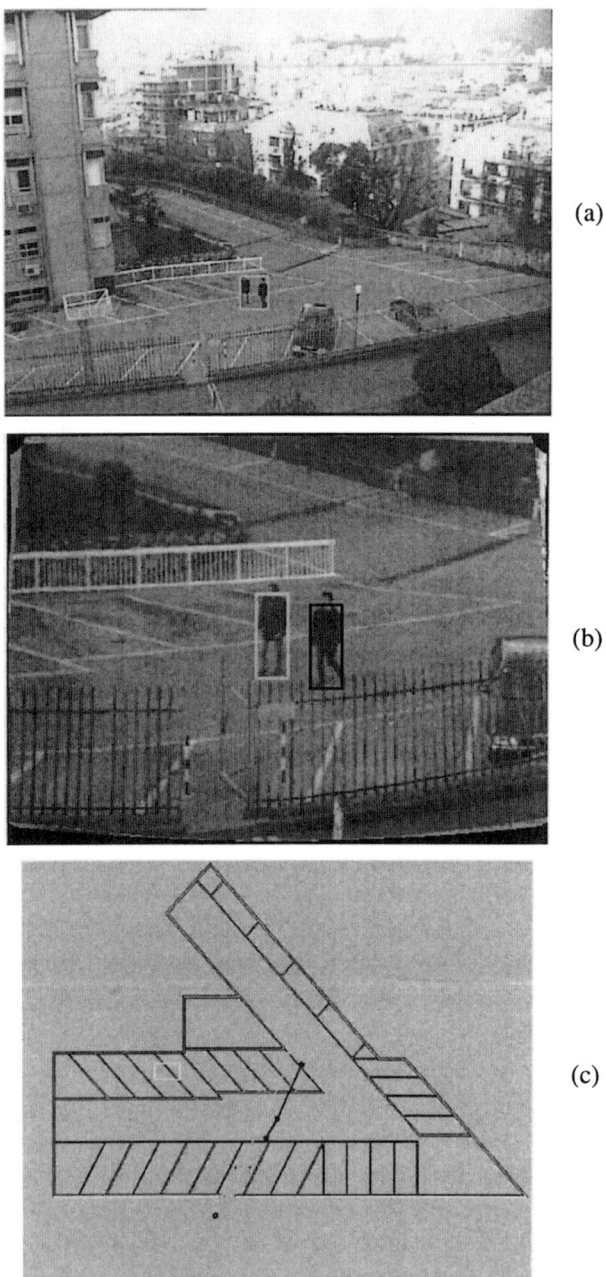

Figure 2-5. A group of people is seen as a single object by the static camera (a) but it is solved by using the mobile sensor (b); trajectory is more precisely estimated by using a zoomed point of view (c)

2. Distributed metadata extraction strategies in a multiresolution dual camera system

Figure 2-6. Automatically generated panoramic background used for change detection by the mobile camera at an intermediate zoom level.

Table 2-1 describes the sequences considered: columns 2 and 3 shows the number of images considered for the three different sequences for static and mobile sensors. Fourth column reports the number of events that are missed by static camera (blobs not correctly detected because of low contrast in the images), fifth column shows the number of false events detected by the mobile subsystem, while the last three columns show the total number of events in each considered sequence.

Table 2-1. A summary of the considered sequences

Seq	Static cam	Mobile cam	Missed static	False mobile	Gate events	New events		Old events	
						car	ped.	car	ped.
Seq1	430	73	25	2	9	6	10	30	16
Seq2	270	52	17	1	8	3	10	5	25
Seq3	154	27	11	0	6	1	3	2	15

Table 2-2. Probabilities of correct detection and error estimates.

Event type	$\hat{P}_d$	$\hat{P}_{err}$
Virtual gate	81.8%	12%
New car detected	90%	-
New pedestrian detected	95%	-
Old car detected	83.3%	-
Old pedestrian detected	94.6%	-

6. CONCLUSIONS

A distributed multi-sensors system able to automatically analyze a complex outdoor environment has been described. In particular the system is able to:
- detect interesting situations within the guarded scene;
- get more precise details of a detected situation allowing a robust scene interpretation
- evaluate moving objects trajectories from two different points of view, thus allowing a more accurate position estimation

The achieved processing time of the system is quite low (about 2 fps) but this is partially justified by considering that the code was not optimized and there are fixed temporal constraints due to the time needed to the pan/tilt unit to get a new position.

Proposed system will be extended to multiple sensors by generalizing concepts presented in this paper; a strategy will be implemented in order to automatically update the panoramic background in order to take into account outdoor illumination variations and other factors that are responsible for a non-static reference frame. Beside this modules for change detection and objects tracking at the intermediate zoom level should be optimized.

ACKNOWLEDGEMENTS

This work was partially supported by the British Council.

REFERENCES

[1] "Multimedia Video-Based Surveillance Systems: Requirements, Issues and Solutions", G.L. Foresti, P. Mahonen and C.S. Regazzoni (Eds.) – Kluwer Academic Publishers, 2000.

[2] *C.S. Regazzoni and G.L. Foresti*, "Video Processing and Communications in Real-Time Surveillance Systems", Real-Time Imaging Journal, Special Issue on Video Processing and Communications in Real-Time Video-Based Surveillance, 2000 (in press).

[3] Multimedia Signal Processing, Special Issue of the Proc. of the IEEE, Part I-II, May-June 1998, Guest Editors T. Chen, K.J. Ray Liu, A.M. Tekalp.

[4] *I. Kuroda and T. Nishitani*, "Multimedia Processors", Proceedings of the IEEE, Vol. 86, No. 6, June 1998.

[5] *V. Morellas, I. Pavlidis, P.T. Siamyrtzis, S. Harp, K. Haigh, M. Bazakos*, "DETER: Detection of Events for Threat Evaluation and Recognition", Proceedings of the

IEEE, Special Issue on Video Communications, Processing and Understanding for Third Generation Surveillance Systems.

[6] *E.Stringa and C.S.Regazzoni*, "Real-Time Video-Shot Detection for Scene Surveillance Applications", IEEE Trans. on Image Processing, USA, Jan 2000, vol 9(1), pp 69-80.

[7] *T.N. Tan, G.D. Sullivan, K.D. Baker,* "Recognizing Objects on the Ground-plane", Image and Vision Computing, 12, No. 3, 164-172, 1994.

[8] *V. Kettnaker and R. Zabih,* "Bayesian Multi-camera Surveillance", Computer Vision and Pattern Recognition, 23 - 25 June 1999, pp. 253-259, Fort Collins, Colorado, USA.

[9] *C.Sacchi, C.S.Regazzoni, C.Dambra*, "Use of video advanced surveillance and communication technologies for remote monitoring of protected sites", Advanced Video-Based Surveillance Systems, C.S. Regazzoni, G. Fabri, G. Vernazza eds., Kluwer Academic Publishers, Norwell, MA, USA, 1999, pp. 154-164.

[10] *R. Mattone, A. Glaeser, and B. Bumann*, "A New Solution Philosophy for Complex Pattern Recognition Problems: Application to Advanced Video-Surveillance", Multimedia Video-Based Surveillance Systems: Requirements, Issues and Solutions, Editors: G.L. Foresti, P. Mahonen and C.S. Regazzoni, Kluwer Academic Publishers, 2000, pp. 94-103.

[11] *B. Peters, J. Meehan, D. Miller, and D. Moore*, "Sensor link protocol: linking sensor systems to the digital battlefield", in Proc. of IEEE Military Communications Conference, Vol. 3 , 1998 , pp. 919 –923.

[12] *M.T. Fennell, and R.P. Wishner*, "Battlefield awareness via synergistic SAR and MTI exploitation", IEEE Aerospace and Electronics Systems Magazine, Vol. 13, No. 2, Feb. 1998, pp. 39-43.

[13] *G.A.Van Sickle*, "Aircraft self reports for military air surveillance", in Proc. of IEEE Digital Avionics Systems Conference, Vol. 2, 1999, pp. 2-8.

[14] *G. Thiel*, "Automatic CCTV Surveillance - Towards The Virtual Guard", IEEE International Carnahan Conference on Security Technology (ICCST) October 5-7, 1999, pp. 42–48, Madrid, Spain.

[15] *L. Marcenaro, F. Oberti and C.S. Regazzoni*, "Change detection methods for automatic scene analysis by using mobile surveillance cameras", Proc. IEEE Int. Conference on Image Processing, Vancouver, Canada, pp. 244 – 247 , 2000.

[16] *L.Marcenaro, F.Oberti, G.L.Foresti and C.S.Regazzoni*, "Distributed architectures and logical task decomposition in multimedia surveillance systems", Proceedings of the IEEE, Vol. 89, N.10, October 2001, pp. 1419-1440.

[17] *L.Marcenaro, F.Oberti and C.S.Regazzoni* "Extending Real Time Change-Detection Techniques To Mosaic Backgrounds And Mobile Camera Sequences In Surveillance Systems", IETE Journal of Research, Special Issue on Visual Media Processing (in press) 2002.

Chapter 3

AUTOMATIC TARGET ACQUISITION AND TRACKING WITH COOPERATIVE FIXED AND PTZ VIDEO CAMERAS

B. Abidi, A. Koschan, S. Kang, M. Mitckes, and M. Abidi

The Imaging, Robotics, and Intelligent Systems Laboratory, The University of Tennessee, Knoxville, 334 Ferris Hall, Knoxville, TN 37996-2001

Key words: Video tracking, active shape models, optical flow, color analysis, mosaicing

1. INTRODUCTION

This chapter presents an overview of an automated video tracking and location system under development at the University of Tennessee's Imaging, Robotics and Intelligent Systems (IRIS) Laboratory in Knoxville, Tennessee. Utilization of this system in any situation where detection of "wrong way" motion with subsequent video tracking would be beneficial is feasible. Examples of these situations include federal buildings, court houses, large office buildings, military bases, and national laboratories. Guidance of a robot arm employing dynamic imaging and motion trajectory analysis of workers in hazardous environments are also potential applications of the system's tracking aspect. The University of Tennessee is initially developing this system for potential use as a security tool in commercial airports.

With the number of people traveling by plane today, security in airport terminals is of great concern. Whenever a suspicious individual is identified or a threat is suspected, the entire section of the airport where the threatening activity is taking place must be cleared for investigation [1]. Knowing or recording the activity of a "bolter" at all times would limit the investigation to a smaller area of the airport and/or even facilitate an effective and risk-favorable apprehension of the violator.

Currently, no systems exist that can automatically and fully perform this task. The system examined in this chapter represents a paradigm shift from the current analog, disconnected, and human intensive security and surveillance systems to a digital, networked, and fully automated system. This system is a camera based security system consisting of a network of cooperating cameras controlled by computer vision software. Automatic target acquisition is performed via cooperating fixed and pan, tilt, zoom (PTZ) cameras, while tracking is achieved solely via PTZ cameras. Several algorithms were proposed in the past to extend camera views to track objects in large areas. Lee et al. proposed a method to align the ground plane across multiple views to build common coordinates for multiple cameras [2]. Dellaert et al. proposed a fast image registration algorithm between the image from a pan/tilt camera and background images from a database [3]. Onmi-directional cameras were also used to extend the field of view to 360° [4]. But the reality remains that most tracking algorithms cater only to the case of fixed cameras and are generally based on adaptive background generation and subtraction [2, 5, 6]. With the system described in this chapter, the constantly changing background with PTZ cameras is a major issue to be addressed and for which a novel background generation methodology is described in section 4.

Another issue facing automatic tracking in public areas is occlusion and features' robustness. Tracking algorithms based on gray level images [7], shape information [8], and color [9] have been proposed before. But despite the various levels of accuracy in modeling objects to be tracked, the assumption must be made in some applications that the object is non-rigid or deformable. In order to represent a non-rigid object (people), active shape models (ASMs) are very efficient compact models in which the shape variety of an object class is taught in a training phase. In this paper, a hierarchical robust approach to an enhanced ASM is proposed to realize an efficient color video tracking system.

Section 2 of this chapter provides information on the automatic detection of breaches. This is followed in section 3 by discussion of automatic target acquisition and handover. Section 4 covers background generation for tracking and is followed by discussion of automatic tracking using PTZ cameras in section 5. Conclusions are presented in section 6.

3. Automatic target acquisition and tracking with cooperative fixed and PTZ video cameras

2. AUTOMATIC DETECTION OF SECURITY BREACHES IN ONE-WAY ACCESS AREAS

A "breach detection system" consisting of a single, fixed, off-the-shelf Sony SSC-DC393 (with auto iris lens) camera was implemented. The system can detect individuals moving against the direction of normal or correct flow and sound an alarm to alert a nearby human operator. This system has been mounted, as a test version, in an actual airport. Figure 3-1 depicts, in (a) a simulation of an exit lane area showing both the prohibited (solid) and the allowed (dashed) directions of motion in this area. In (b) the actual one camera test system is shown.

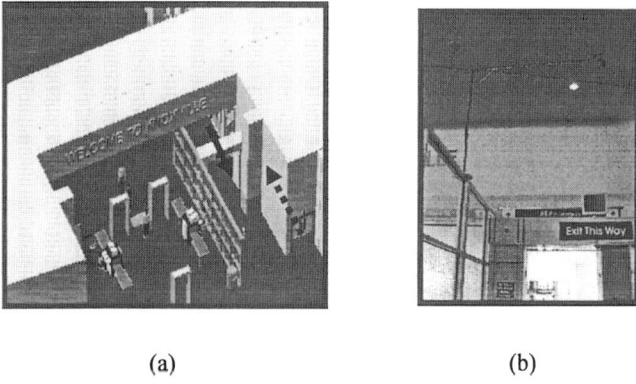

(a)   (b)

Figure 3-1. Breach detection system – (left) simulation of exit lane, (right) actual system being tested

In-lab and field experiments were conducted and access breaches detected and color-coded on the monitor with subsequent alarm activation. Figure 3-2 illustrates the detection of an access breach (enclosed in the rectangular shape) in (a) and an overall screen view in (b).

Optical flow methods were used to compute motion vectors based on the intensity values of successive image frames. Figure 3-3 depicts the basic steps of the overall breach detection and tracking algorithm.

The assumption is that the intensity values of an object do not vary when the object is moving in space. If $S_c(x_1, x_2, t)$ is the continuous space-time intensity distribution, then the following equation can be used

$$\frac{dS_c(x_1, x_2, t)}{dt} = 0. \tag{3.1}$$

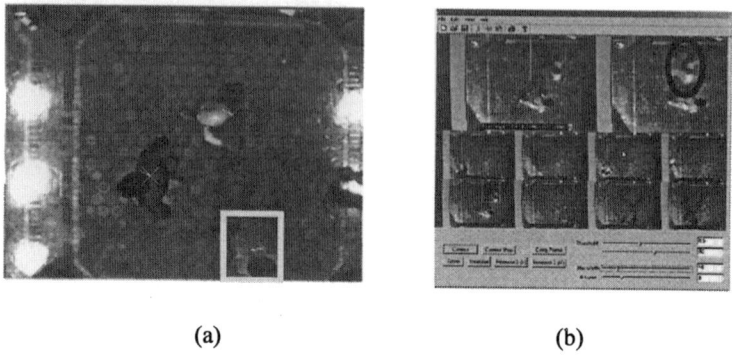

Figure 3-2. Breach detection from a single fixed overhead camera – (a) close-up view, (b) GUI designed to display the breach occurrence and detection

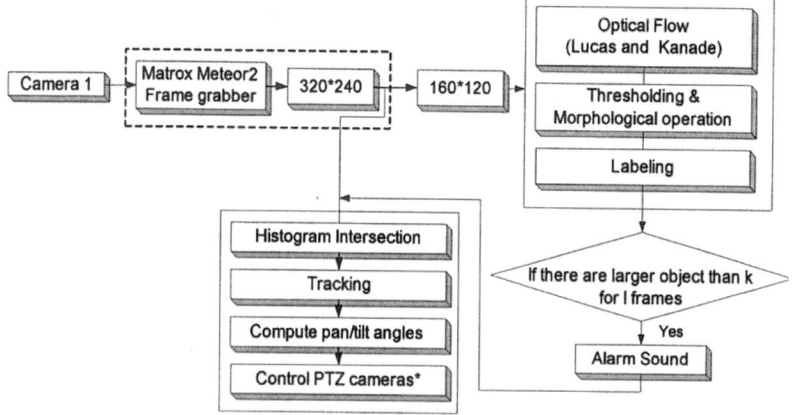

Figure 3-3. Flow chart of the breach detection and tracking system

Equation 3.2 can be formulated with the chain rule of differentiation as

$$\frac{\partial S_c(\mathbf{x};t)}{\partial x_1}v_1(\mathbf{x};t) + \frac{\partial S_c(\mathbf{x};t)}{\partial x_2}v_2(\mathbf{x};t) + \frac{\partial S_c(\mathbf{x};t)}{\partial t} = 0. \quad (3.2)$$

The optical flow equation can then be expressed as

$$\varepsilon_{of}(\mathbf{v}(\mathbf{x},t)) = <\nabla S_c(\mathbf{x};t), \mathbf{v}(\mathbf{x},t)> + \frac{\partial S_c(\mathbf{x};t)}{\partial t}. \quad (3.3)$$

The displacement **x** is the quantity found by minimizing Equation (3.3). This equation, however, uses the derivative of neighboring pixels, and is therefore sensitive to noise. Two different approaches widely used in

3. Automatic target acquisition and tracking with cooperative fixed and PTZ video cameras

literature to reduce noise are the Horn and Schunck's method [10] and the Lucas and Kanade's approach. Lucas and Kanade proposed a block motion model [11], where the assumption is that the motion vector remains unchanged over a particular block of pixels. The error can then be defined as

$$E = \sum_{x \in Neighbor} (\varepsilon_{of})^2 \qquad (3.4)$$

and the solution, which minimizes the error in a block, can be formulated as

$$\begin{bmatrix} \hat{v}_1 \\ \hat{v}_2 \end{bmatrix} = \begin{bmatrix} \sum_{x \in Neighbor} \frac{\partial S_c(x;t)}{\partial x_1} \frac{\partial S_c(x;t)}{\partial x_1} & \sum_{x \in Neighbor} \frac{\partial S_c(x;t)}{\partial x_1} \frac{\partial S_c(x;t)}{\partial x_2} \\ \sum_{x \in Neighbor} \frac{\partial S_c(x;t)}{\partial x_1} \frac{\partial S_c(x;t)}{\partial x_2} & \sum_{x \in Neighbor} \frac{\partial S_c(x;t)}{\partial x_2} \frac{\partial S_c(x;t)}{\partial x_2} \end{bmatrix}^{-1} \begin{bmatrix} -\sum_{x \in Neighbor} \frac{\partial S_c(x;t)}{\partial x_1} \frac{\partial S_c(x;t)}{\partial t} \\ -\sum_{x \in Neighbor} \frac{\partial S_c(x;t)}{\partial x_2} \frac{\partial S_c(x;t)}{\partial t} \end{bmatrix} \qquad (3.5)$$

Once the motion vectors were computed, the regions are segmented and labeled to depict the wrong way motion.

3. AUTOMATIC TARGET ACQUISITION AND HANDOVER FROM FIXED TO PTZ CAMERA

When a breach occurrence is detected, the fixed camera in charge of monitoring the direction of motion triggers an alarm and provides the position of the target in the world coordinate system. The PTZ, Panasonic WV-CS854A camera then uses that position information to determine its pan and tilt angles and lock on the target for subsequent tracking. Figure 3-4 depicts a simulated view of the overhead fixed and front PTZ camera system in (a) and the geometry of the system in (b). The pan and tilt angles for the PTZ camera are respectively given in equation (3.6) as a function of the coordinates (x_t, y_t, h_t) of the target

$$\theta = \sin^{-1} \frac{x_t}{\sqrt{x_t^2 + y_t^2}}, \quad \delta = \cos^{-1} \frac{\sqrt{x_t^2 + y_t^2}}{\sqrt{x_t^2 + y_t^2 + (h_c - h_t)^2}}. \qquad (3.6)$$

Handover is only considered complete when the PTZ camera is able to extract the moving target from its background and lock on it. This step is achieved using the same principle of direction of motion; only this time the motion being searched for is top down motion instead of left right as illustrated in Figure 3-5. A GUI view of a typical image captured from the two-camera system is shown in Figure 3-6, whereas Figure 3-7 shows successive target views from the PTZ camera. Two Matrox Meteor2 frame grabbers and a Pentium 4@1.4GHz were used in this target capturing and tracking procedure.

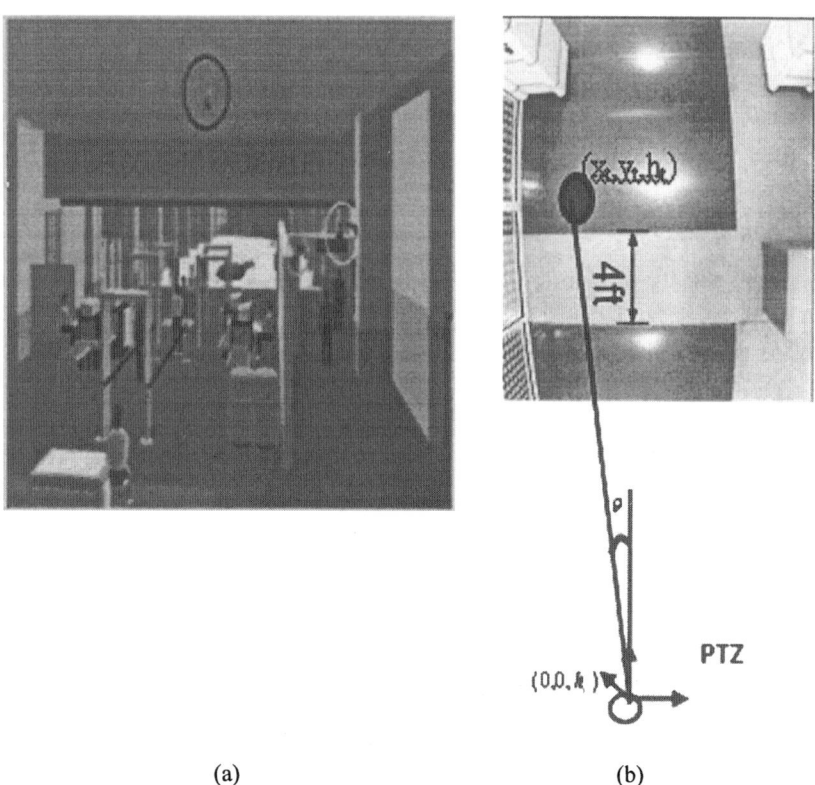

(a)   (b)

Figure 3-4. (a) Simulated view of the multi-camera system for automatic target acquisition and tracking, (b) geometry of the dual-camera system

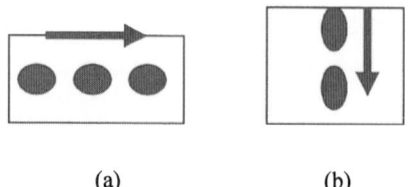

(a)   (b)

Figure 3-5. Use of direction of motion for automatic target acquisition and handover: (a) simulated video from overhead camera with left to right motion and (b) video from PTZ front camera with top-down motion

3. Automatic target acquisition and tracking with cooperative fixed and PTZ video cameras

4. BACKGROUND GENERATION FOR TRACKING WITH PTZ CAMERAS

Phase 3 in the automatic target acquisition and tracking system involves subject tracking with the PTZ. One of the most challenging aspects in PTZ tracking is the background generation since both the camera and the target are moving at the same time. Following we propose a background modeling scheme for PTZ cameras that deals with relative motion between the camera and the target. An adaptive background generation procedure for fixed cameras and its application using PTZ cameras is shown in Figure 3-8. When a single, fixed camera is used, each location of a stationary pixel, denoted by the dark rectangles in Figure 3-8(a), is time-invariant. Figure 3-8(b) illustrates how the location of the dark rectangles changes depending on the camera's motion of tilting and panning. The rectified images in terms of

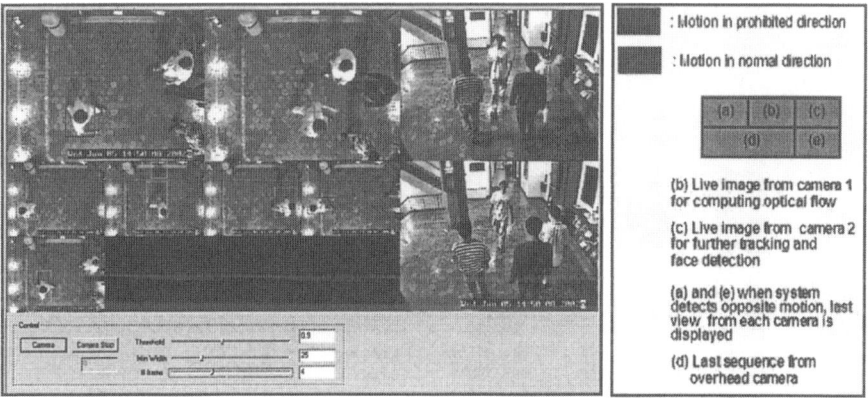

Figure 3-6. GUI for the two camera system

Figure 3-7. Sequence of frames from PTZ camera showing achieved handover

position of corresponding points are shown in Figure 3-8(c). Mosaicing enables a generated background for one viewpoint to be reused as the background for another view if there is an overlap region. This overlapping condition holds in most practical situations of tracking with a PTZ camera.

Two methods were considered for the registration of images with different pan and tilt angles. The first uses the 8-points algorithm and the other uses the pan/tilt angles and the focal length. The 8-points algorithm requires at least 4 corresponding points. If we can change the pan and tilt angles by the same amount, then we always get the same corresponding locations for a pair of images. The first task is to compute the suitable pan and tilt angles for tracking. These angles should be selected to guarantee at least 50% overlap depending on the zoom ratio. After selecting these angles,

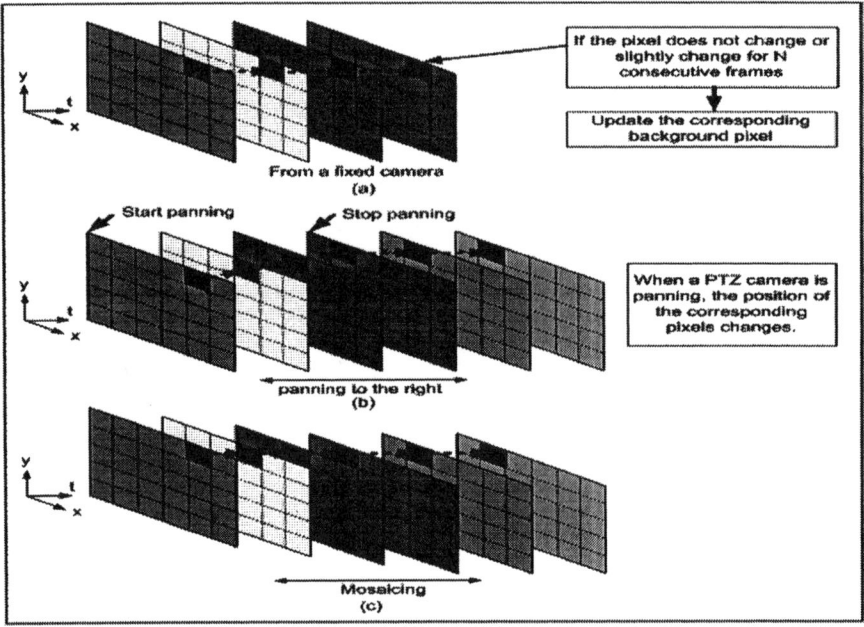

Figure 3-8. Illustration of the background changes with PTZ cameras: (a) images captured by a fixed camera, (b) images captured by a panning camera, and (c) the rectified version of (b) in terms of each pixel's location

4 or more corresponding points need to be picked to compute the homogeneous matrix, which describes the relation between a pair of images with different pan and tilt angles. This process is illustrated in Figure 3-9. The 4 corresponding points are denoted by F1 to F4 in Figure 3-9(a). Once the homogeneous matrix is computed, the generated background is projected

3. Automatic target acquisition and tracking with cooperative fixed and PTZ video cameras

for the new view using this matrix. The advantage of this method is that there is no need to know the internal parameters of the PTZ camera, such as the size of the CCD sensor and the focal length, but the homogeneous matrices for every possible combination of pan and tilt angles have to be computed beforehand.

Instead of computing the homogeneous matrices, we can register a pair of images by 3D rotation if the camera's internal parameters, such as the actual CCD size and the focal length, are known. Two characteristics of PTZ cameras make this method possible; 1) the optical center is always perpendicular to the CCD sensor and this feature is invariant to panning and tilting and 2) the distance between the optical center and the center of the image is fixed for the same zoom ratio.

The first step for this task is to express each 2D point (x, y) by a 3D unit vector as function of angles δ, θ. The center of the image is $\delta = 0, \theta = 0$. The two angles δ, θ change by either the zooming ratio or the values of (x, y).

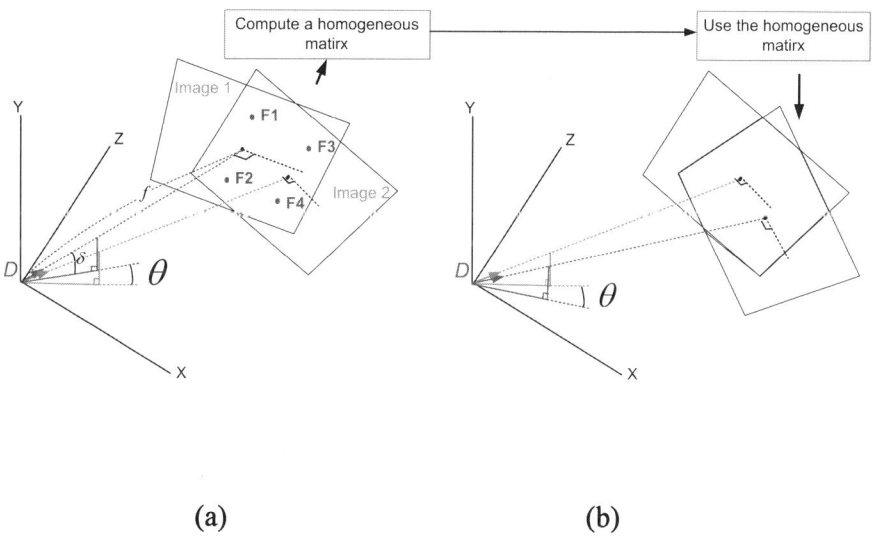

(a)   (b)

Figure 3-9. Registration of a pair of images: (a) a pair of images with the same tilt angles and different pan angles are used to compute the homogeneous matrix, and (b) the previously computed homogeneous matrix is used to register a new image with a different pan angle

For instance, F1 in Figure 3-9a can be expressed by not only two actual locations (x, y) for both images 1 and 2, but also by a set of δ, θ in the world coordinate system.

The second step is to perform 3D rotations along the X- and Y-axes. Let image 2 be the new image after changing the pan and tilt angles. Then the unit vector for point F1 in image 2 needs to be transformed to find the corresponding location in image 1. The first rotation along the X horizontal axis is performed to compensate for the tilt angle of image 2, so the rotation angle is − (the tilt angle of image 2). The second rotation along the vertical Y axis compensates for the difference in panning angles, which is (the panning angle for image 2 − the panning angle for image 1), and the third compensates for the tilt angle of image 1. After these operations, we can get a 3D rotationally transformed unit vector, and the matching point can be computed by converting δ, θ to 2D (x, y) for image 1. The final result is image 1 transformed into image 2.

Figure 3-10 illustrates the transformed background shown in the first row. Mosaicing is accomplished in the first two columns and then updating is performed. In the second row are the images from the PTZ camera, and the third row shows the detected moving regions as white pixels on a black, stationary, background. Since we did not generate a new background for the current position, error in the motion detection process shows up as motion in objects that are obviously stationary. This can be resolved by generating a background for the new position in advance and combining that background with the background transformed from the previous frame as in column 5 of Figure 3-10.

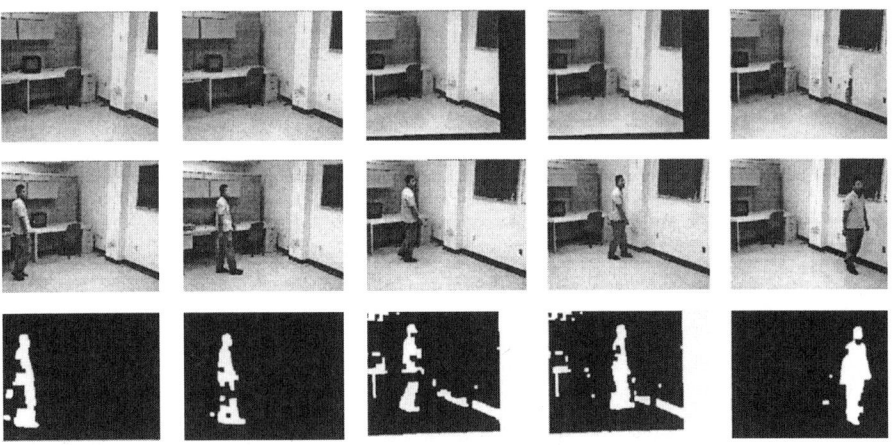

Figure 3-10. Background generation for PTZ camera using mosaicing; background images (top), PTZ view (middle), and motion detected (bottom)

5. AUTOMATIC TRACKING VIA PTZ CAMERAS

Two different approaches for tracking people in video sequences are presented.

5.1 Color and Predicted Direction and Speed of Motion

Image distortions caused by PTZ cameras make the tracking task difficult. Features that are robust to these distortions are needed for the tracking task. Color information of the target can be such a feature. When color constancy is preserved, the color distribution of interesting regions can be used to track objects. Color indexing [12] is one of the techniques used to find similar color targets in consecutive frames. The video from the overhead camera is first analyzed to detect and extract breaches. Each extracted region is used to build a color histogram model. Once the histogram models are acquired, the nearest and most similar color regions are searched through histogram intersection. The results are trajectories of the objects that caused the alarm. Experimental results using the histogram intersection are shown in Figure 3-11. Since the trajectories were computed for each frame, the speed and direction of motion can also be predicted and used to compute the internal parameters of the PTZ camera, such as pan and tilt angles. The PTZ camera is then automatically controlled to view the predicted location and to extract the top-down motion caused by the breach. A verification process will then follow to check if the extracted regions are effectively caused by the breach.

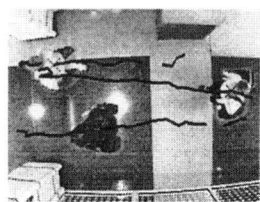

Figure 3-11. Tracking results using color indexing

Another promising algorithm for the tracking and recognition of individuals in video image sequences that is robust to occlusion is based on color ASMs and is presented in the following subsection.

5.2 Color Active Shape Models

Active shape models can be applied to the tracking of people. The shape of a human body has a unique combination of head, torso, and legs, which can be modeled with only a few parameters of the ASM. ASM-based video tracking can be performed in the following order: (a) shape variation modeling, (b) model fitting, (c) local structure modeling, and (d) in our approach, an additional color component analysis.

Given a frame of input video, suitable landmark points should be assigned on the contour of the object. Good landmark points should be at the same location for each shape. In a two-dimensional image, we represent n landmark points by a $2n$-dimensional vector as $\mathbf{x} = [x_1,...,x_n, y_1,...,y_n]^T$.

A set of n landmark points represents the shape of an object as shown in Figure 3-12. A set of frames can make a training set. Although each shape in the training set is in the $2n$-dimensional space, we can model the shape with a reduced number of parameters using Principal Component Analysis (PCA).

The best pose and shape parameters to match a shape in the model coordinate frame, $\mathbf{x}$, to a new shape in the image coordinate frame, $\mathbf{y}$, can be found by minimizing the following error function

$$E = (\mathbf{y} - \mathbf{Mx})^T \mathbf{W}(\mathbf{y} - \mathbf{Mx}), \qquad (3.7)$$

where $\mathbf{M}$ represents the geometric transformation of rotation θ, translation $\mathbf{t}$, and scale s. After the set of pose parameters, $\{\theta, \mathbf{t}, s\}$, is obtained, the projection of $\mathbf{y}$ into the model coordinate frame is given as $\mathbf{x}_p = \mathbf{M}^{-1}\mathbf{y}$. The model parameters are updated as $\mathbf{b} = \mathbf{\Phi}^T(\mathbf{x}_p - \mathbf{x})$.

A statistical, deformable shape model can be built by landmark point's assignment, PCA, and model fitting steps. In order to interpret a given shape in the input image based on the shape model, we must find the set of parameters that best matches the model to the image.

If we assume that the shape model represents boundaries and strong edges of the object, a profile across each landmark point has an edge-like local structure. Let $\mathbf{g}_j$, $j=1,...,n$, be the normalized derivative of a local profile of length K across the j-th landmark point, and $\overline{\mathbf{g}}_j$ and $\mathbf{S}_j$ the corresponding mean and covariance, respectively. The nearest profile can be obtained by minimizing the following Mahalanobis distance between the sample and the mean of the model as

3. Automatic target acquisition and tracking with cooperative fixed and PTZ video cameras

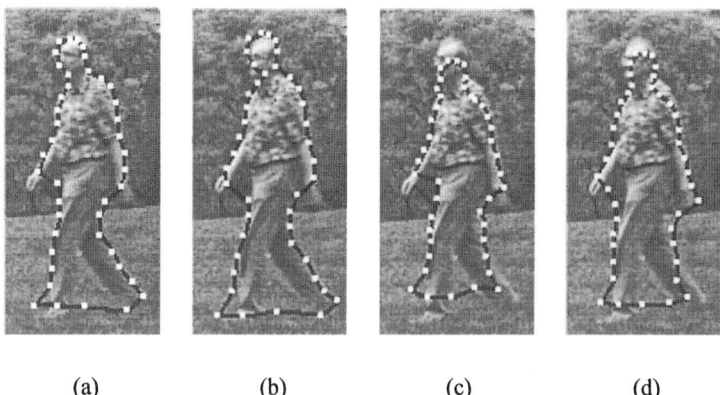

(a) (b) (c) (d)

Figure 3-12. Fitting results for the 4th frame of the Man_9 sequence using the hierarchical method with the median selection mode in the (a) intensity, (b) *RGB*, (c) *HSI*, and (d) *YUV* spaces

$$f(\mathbf{g}_{j,m}) = (\mathbf{g}_{j,m} - \overline{\mathbf{g}}_j)^T \mathbf{S}_j^{-1} (\mathbf{g}_{j,m} - \overline{\mathbf{g}}_j), \qquad (3.8)$$

where $\mathbf{g}_{j,m}$ represents $\mathbf{g}_j$ shifted by m samples along the normal direction of the corresponding boundary.

In gray level image processing, the objective functions are determined along the normal vectors for representative points in the gray value distribution. This procedure can be extended to color images by first computing the objective functions separately for each component of the color vectors. Afterwards, a "common" minimum has to be determined by analyzing the resulting minima that are computed for each single color component. One means of doing this consists of selecting the absolute minimum in the three color components as a candidate. Another procedure consists of selecting the average of the absolute minima in all three color components. However, outliers in one color channel lead in both cases to the wrong result. One way to overcome this problem is to use the median of the absolute minima in the three color channels as a candidate. Thereby the influence of outliers in the minima of the objective functions is minimized.

We studied the performance of the ASM when employing the color spaces *RGB*, *YUV* and *HSI*. So far we have applied the same procedure to all color spaces. In our experiments, we obtained the best results when using the median in the *RGB* space (see Figure 3-12 and Table 3-1). In addition, we applied a hierarchical implementation using image pyramids to speed up the process and decrease the error [13].

Table 3-1. Error between the manually assigned points and the estimated points using three different minimum selection methods in different color spaces for a selected frame

Color Space	Minimum	Median	Mean
Intensity		164.32	
RGB	208.27	142.29	142.29
YUV	406.78	353.77	196.88
HSI	251.68	343.74	207.36

5.3 Occlusions and Illumination Changes

One advantage of ASM-based tracking is its ability to follow the shape of an occluded object. We studied outdoor sequences in the *RGB* color space, where individuals are partially occluded by different objects. Results obtained when applying the hierarchical method with the median selection mode to the sequence Man_11 are shown in Figure 3-13. The proposed tracking scheme provided good results in our experiments, even though the object is partially occluded by a bench. One property of the ASM-based tracking scheme is that the ASM can easily adjust to reappearing parts of the tracked object in an image sequence.

Tracking of a person becomes rather difficult if the image sequence contains several, similarly shaped moving people. In this case, a technique based exclusively on the contour of a person will have difficulties in tracking a selected individual. On the other hand, a technique exclusively evaluating the colors of a moving person (or object) may also fail. Any color-based tracker can lose the object it is tracking due, for example, to occlusion or changing lighting conditions. To overcome the sensitivity of a color-based tracker to changing lighting conditions, the color constancy problem has to be solved at least in part, which is a non-trivial and a computationally costly task.

A possible solution to this problem might consist of a weighted combination of an ASM form-based and a color indexing tracking technique. By applying such a combination technique to image sequences we might be able to distinguish between: a) objects of similar colors but with different forms, and b) objects of different colors but with similar forms. One drawback of such a combination approach is its high computational cost. Here a hardware implementation can be considered later for real-time applications.

3. Automatic target acquisition and tracking with cooperative fixed and PTZ video cameras

Figure 3-13. Fitting results in two frames of a video sequence with a partially occluded person. The hierarchical method with the median selection mode in the *RGB* color space was used

6. CONCLUSION

An automatic breach detection, target acquisition, and tracking system was designed. The system is based on the cooperative work of a single fixed and a network of PTZ cameras. The breach detection system consists of a single camera and an optical flow algorithm serves for the acquisition of the target. The target is then handed over to a PTZ camera for tracking and location reporting. The handover is achieved through geometric modeling and top-down motion detection. A novel background generation method based on projective geometry is then used to extract motion from a PTZ camera signal. Two tracking techniques were implemented, one employs color indexing and the other is an active shape modeling approach, which was determined to deal very well with occlusion.

Future research and development will address the fine tuning of this system. Detection of "running people" through controlled access areas will be implemented using one of two possible approaches: a hierarchical approach with Gaussian image pyramids or a high-frame rate camera. Noise removal for the breach detection system can be addressed either by using a high-quality camera or an adaptive background generation and subtraction. Furthermore, the active shape modeling technique has to be optimized to run in real time and for use in tracking in cluttered environments. A fusion of the two tracking methods will also be tested.

ACKNOWLEDGMENT

This work was supported by the TSA/NSSA Program, R01-1344-49, the University Research Program in Robotics under grant DOE-DE-FG02-86NE37968, and by the DOD/TACOM/NAC/ARC Program, R01-1344-18.

REFERENCES

[1] B. Abidi, D. Shelton, M. Mitckes, J. Paik, and M. Abidi, "Gate-to-Gate Automated Video Tracking / Location- End of Year Report 07/01/2000-06/30/2001," The IRIS lab, UTK, September 2001.

[2] L. Lee, R. Romano, and G. Stein, "Monitoring activities from multiple video streams: establishing a common coordinate frame," IEEE Trans. On Pattern Analysis and Machine Intelligence, Vol. 22, No. 8, pp. 758-767, August 2000.

[3] F. Dellaert and R. Collins, "Fast image-based tracking by selective pixel integration," ICCV 99 Workshop on Frame-Rate Vision, September 1999.

[4] M. Nicolescu, G. Medioni, and M. Lee, "Segmentation, tracking and interpretation using panoramic video," IEEE Workshop on Omnidirectional Vision, pp. 169-174, 2000.

[5] I. Haritaoglu, D. Harwood, and L. S. Davis, "W4: Real-time surveillance of people and their activities," IEEE Trans. On Pattern Analysis and Machine Intelligence, Vol. 22, No. 8, pp. 809-830, August 2000.

[6] T. Horprasert, D. Harwood, and L.S. Davis, "A robust background subtraction and shadow detection," Proc. ACCV 2000, Taipei, Taiwan, January 2000.

[7] R. Plankers and P. Fua, "Tracking and modeling people in video sequences," Computer Vision and Image Understanding, Vol. 81, pp. 285-302, 2001.

[8] A. Blake and M. Isard, Active Contours, Springer, London, England, 1998.

[9] S. J. McKenna, Y. Raja, and S. Gong, "Tracking colour objects using adaptive mixture models," Image and Vision Computing, Vol. 17, pp. 225-231, 1999.

[10] B.K.P. Horn and B.G. Schunck, "Determining optical flow," Artificial Intelligence, Vol. 16, pp. 185-203, August 1981.

[11] B. D. Lucas and T. Kanade, "An iterative image registration technique with an application to stereo vision," Proc. DARPA Image Understanding Workshop, pp. 121-130, 1981.

[12] M. J. Swain and D. H. Ballard, "Color indexing," International Journal of Computer Vision, pp. 11-32, 1991.

[13] S. K. Kang, H. S. Zhang, J. K. Paik, A. Koschan, B. Abidi, and M. A. Abidi, "Hierarchical approach to enhanced active shape model for color video tracking," Proc. Int. Conf. on Image Processing ICIP02, Rochester, N.Y., Vol. I, pp. 888-891, 2002.

Chapter 4

LEARNING THE FUSION OF VIDEO DATA STREAMS
Automatic Calibration and Registration of Surveillance Cameras

J.Renno, P.Remagnino and G.A.Jones
Digital Imaging Research Centre, Kingston University, Penrhyn Road, Kingston Upon Thames, Surrey, UK, KT1 2EE.

Key words: Visual Surveillance, Machine Learning, Data Fusion, Camera Calibration.

1. INTRODUCTION

Visual surveillance cameras pose two major problems: on one hand they must be installed and calibrated, on the other hand camera views must be combined if the final goal is to generate some natural language description of the dynamic evolution of a scene[3][4][5][7][8][9]. The work presented in this paper provides a semi automated solution to both problems.

Camera calibration with single and multiple cameras is at least a decade old (see [14][15][6] for in depth description of the field), many solutions have been proposed and a review would be outside the scope of this paper. The fusion of video data, on the other hand, is relatively new: two examples are the work of Chang [12] and Black [13].

The usual technique involves the calibration of all deployed cameras using a set of points in the scene, either lying on a plane or scattered in a 3D space. The idea is to create a common Cartesian reference frame for all the cameras and then back-project relevant events (image bounding boxes or blobs) onto the defined frame, integrating all video information [2]. What this paper proposes is the development of an automatic way of both calibrating and registering surveillance cameras, based merely on the observation of activity from two or more cameras.

First we develop highly discriminatory bounding-box appearance models of scene objects which indirectly use the depth of the object to model its projected width and height. Since, the spatial extent of an object is now a function of image position, the tracker will be more robust when presented with the distorted observations which arise from fragmentation or occlusion processes. Second, the observations are transformed onto the ground plane coordinate system within which a quadratic rather than linear motion model is defined. Global real-valued noise models can be generated for observation and dynamic noise models. Finally, rather than relying on a labour-intensive calibration procedures to recover the image to ground-plane homography [11], the system relies on a simple auto-calibration procedure to learn the relationship between image and world by simply watching events within the monitored scene. Having calibrated each camera to its local ground plane, Section 4 demonstrates how ground planes may be registered. Again a learning procedure is pursued in which the projected trajectory positions in corresponding frames and their instantaneous velocity estimates are combined to create estimates of the rotation and translation. A clustering algorithm is used to locate the most likely transform between each pair of camera ground planes.

2. AUTO-CALIBRATION OF THE GROUND PLANE

In this section a simple yet highly effective method of learning the image to ground plane homography of the camera is presented which exploits the simple but reasonably accurate assumption that in typical surveillance installations, *the projected 2D image height of an object varies linearly with its vertical position in the image* - increasing down the image from zero at the horizon. This height model is derived from the optical geometry of a typical visual surveillance installation. In addition, such an assumption enables the use of simple but highly discriminatory models of the appearance of scene objects which indirectly use the depth of the object to model its projected height. In this *auto-calibration* scenario, the *ground plane coordinate system* (GPCS) is defined as follows:

The Y-axis $\hat{Y}$ of the GPCS is defined as the projection of the optical axis along the ground plane. The Z-axis $\hat{Z}$ is defined as the ground plane normal. The position of the camera focal point in the GPCS is `above' the GPCS origin at the point $(0,0,L)$.

2.1 Ground Plane Projection

The image plane is situated at distance f (focal length of the optical system for the camera) perpendicular to the optical axis $\hat{z}$ (see Figure 4.1(a)). In this configuration a point P on the image plane has coordinates $x' = (x, y, -f)^T$. The pixel coordinate system i, j (representing the row and column position respectively) is related to the image plane coordinate system by $x = \alpha_x(j - j_0)$ and $y = \alpha_y(i_0 - i)$ where i_0, j_0 is the optical centre of the image and α_x and α_y are the horizontal and vertical inter-pixel widths. Thus $x' = \alpha_x^f(j - j_0), \alpha_y^f(i - i_0), -1)^T f$ where α_x and α_y are the horizontal and vertical pixel dimensions normalised by the focal length.

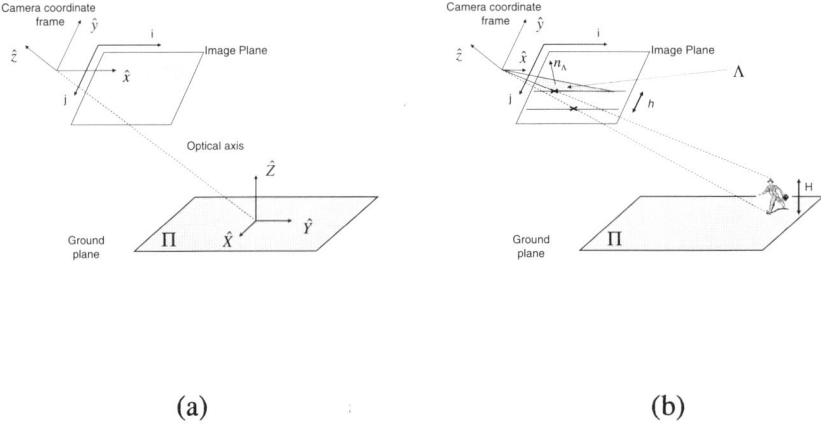

(a)   (b)

Figure 4-1. (a) Camera, World and Pixel Coordinate Systems. (b) Projected Height

An optical ray containing the focal point of the camera passing through the image plane can be represented in vector form as $x = \mu x'$. Let Q be the point of intersection of the optical ray with the ground plane Π. In order to calculate the position of the point Q on the ground plane Π in the ground plane coordinate system, one must convert the position of a point given the transformation (R, t) between the image plane and world coordinate systems i.e. $X = \mu R x' + t$. Writing the ground plane equation as $n_\Pi \cdot X = 0$, where the ground plane normal $n_\Pi \equiv \hat{Z}$, then the position X of the point Q is obtained by noting that $X \cdot \hat{Z} = 0$.

$$\mu = -t_z / \hat{Z} \cdot Rx' \qquad (4.1)$$

The local GPCS is defined with a zero pan angle. Assuming no significant roll angle, then after some algebraic manipulation, the ground plane coordinates may be related to the look-down angle θ as follows

$$\frac{X}{L} = \frac{\alpha_x^f (j - j_0)}{\alpha_y^f (i - i_0) \sin \theta - \cos \theta}, \quad \frac{Y}{L} = -\frac{\alpha_y^f (i - i_0) \cos \theta - \sin \theta}{\alpha_y^f (i - i_0) \sin \theta - \cos \theta} \qquad (4.2)$$

Thus to compute the ground plane position of an image point, the following camera parameters $i_0, j_0, \alpha_x^f, \alpha_y^f$, and θ are needed. In our approach the optical centre i_0, j_0 is computed by an optical flow algorithm which robustly fits a global zoom motion model to a three frame sequence undergoing a small zoom motion.

2.2 Projected Object Height

If one assumes that the height of a moving object is known (*i.e.* a person) then the point of intersection X with the ground plane can be shifted along the $\hat{Z}$ direction by the height H. Using μ, we can write $X' = \mu Rx' + t + H\hat{Z}$. The new image point x'' corresponding to the projection of the top of the person can be computed from the inverse transformation $R^T(X' - t)$ to yield

$$\lambda x'' = \mu x' + H R^T \hat{Z} \qquad (4.3)$$

where λ is the projection factor from the image plane to the top of the person. Substituting μ from Equation (4.1) and $t_z = L$ yields

$$x'' = -\frac{1}{\lambda} \left(H R^T \hat{Z} - \frac{L}{\hat{Z} \cdot Rx'} x' \right) \qquad (4.4)$$

4. Learning the fusion of Video Data streams

To measure the projected vertical height of an object, we simply define a plane Λ containing the optical centre and the image plane raster line containing the new point x'' (see Figure 4.1(b)). The normal n_Λ of this plane is defined by the cross-product between the projection line $\lambda x''$ and the raster line direction vector $\hat{x}$ as follows

$$n_\Lambda = -\frac{1}{\lambda}\left(H\,R^T \hat{Z} \times \hat{x} - \frac{L}{\hat{Z}\cdot Rx'}x' \times \hat{x}\right) \tag{4.5}$$

The raster line containing the point x'' can be thought of as lying at a distance h above the projection of the bottom of the person - see Figure 4-1(b). Therefore the point vertically above x' can be expressed as $x = x' + h\hat{y}$ and belongs to the plane Λ. Substituting $x' + h\hat{y}$ into the equation of plane Λ, $n_\Lambda \cdot x = 0$ generates

$$h = -\frac{n_\Lambda \cdot x'}{n_\Lambda \cdot \hat{y}} \tag{4.6}$$

Further simplification can be derived by expanding the numerator and denominator of Equation (4.6) using Equation (4.5) as follows

$$-\lambda n_\Lambda \cdot x' = H\left(R^T \hat{Z} \times \hat{x}\right)\cdot x' - \frac{L}{\hat{Z}\cdot Rx'}(x' \times \hat{x})\cdot x' \tag{4.7}$$
$$= H\left(R^T \hat{Z} \times \hat{x}\right)\cdot x' \tag{4.8}$$

since $(x' \times \hat{x}) \cdot x' = 0$, and

$$-\lambda n_\Lambda \cdot \hat{y} = H\left(R^T \hat{Z} \times \hat{x}\right)\cdot \hat{y} - \frac{L}{\hat{Z}\cdot Rx'}(x' \times \hat{x})\cdot \hat{y}$$
$$= H\left(R^T \hat{Z} \times \hat{x}\right)\cdot \hat{y} - \frac{Lf}{\hat{Z}\cdot Rx'} \tag{4.9}$$

where $(x' \times \hat{x}) \cdot \hat{y} = f$. Where there is a zero roll angle, Equations (4.8) and (4.9) combine to generate the following expression for image plane height h which depends only on object height H, camera height L and vertical image height y.

$$h = \frac{(f^2 - y^2)\sin\theta\cos\theta + yf(\cos^2\theta - \sin^2\theta)}{y\sin\theta\cos\theta - (\cos^2\theta - L/H)f} \tag{4.10}$$

For typical camera installations, h can be shown to effectively vary linearly with vertical image position relative to the position of horizon. The intercept with the vertical position axis (or $h = 0$ axis) defines the horizon where objects become infinitely small. Such a linear model may be extracted from moving regions of the monitored scene - see Figure 4.2[1] and 4.4. Currently the operator drags a line segment along the ridge structure to define the gradient γ and horizon i_h.

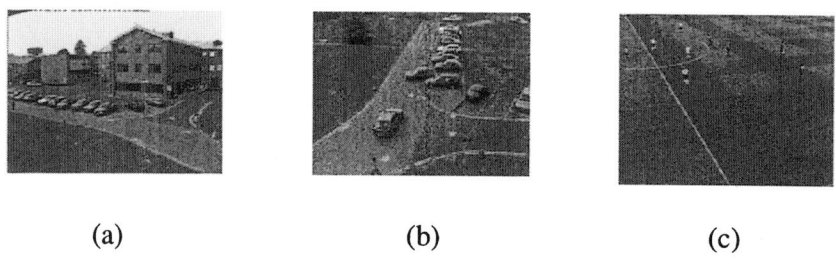

Figure 4-2. (a) PETS Camera 1, (b) PETS Camera 2, (c) Football.

2.3 Ground Plane Calibration

Since the vertical image height of an object is independent of the horizontal image position of the projected object, the following derivation may assume, without loss of generality, that the object is located on the vertical axis *i.e.* $x = 0$. Two key positions of a projected object may be defined at $i = i_h$ at the horizon, and $i = i_0$ at the optical centre of the image. At the former, the look-down angle θ may directly related to the horizon parameter i_h extracted from the accumulated training data acquired in the learning stage described in Section 2.2 *i.e.*

$$\cot\theta = \alpha_y^f (i_h - i_0) \qquad (4.11)$$

For the latter case, consider an object of height H standing on the ground plane point given by the projection of the optical axis. From Equation (4.10), the vertical height at this point $h(i = i_0)$ may be related to the look-down angle as follows

$$\frac{h}{f} = \frac{H\cos\theta\sin\theta}{L - H\cos^2\theta} \qquad (4.12)$$

An estimate of the height h may also be generated from the learnt linear projected height model i.e. $h(i_0) = \alpha_x \gamma (i_0 - i_h)$. Combining this with Equations (4.11) and (4.12), the following expressions for the camera parameters θ and α_y^f may be derived

$$\sin^2\theta = \frac{\gamma}{H}\frac{L-H}{1-\gamma}, \quad \alpha_y^f = \frac{\cot\theta}{(i_0 - i_h)} \qquad (4.13)$$

3. MODEL-BASED TRACKING

In this section, the projected height concept is employed to define simple yet highly effective bounding box appearance models for the principle object types within a surveillance scene. The representation is composed of two vertically adjacent connected bounding boxes - the *object component* and *base component*. The *base* is the large number of background pixels beneath an object and the shadow regions which are typically segmented with the object pixels themselves. The object component is defined by (i) the vertical extent of the object - the *height model*, (ii) the horizontal extent of the object - the *width model*, and (iii) the vertical extent of the base region - the *base model*. These models, as illustrated in Figure 4.3, are defined relative to the 2D position of the object - the 2D projection of the position of the object on the ground plane.

Three different models are currently used corresponding to each of the principle vehicles types Ψ in the set

$\Psi = \{Person, Vehicle, Large\ Vehicle\}$ As with the ground-plane auto-calibration, the parameters for each of these models must be computed in a learning procedure.

(a) (b) (c)

Figure 4-3. (a) Typical object (b) Detected pixels (c) Object model.

$$\mu = \Gamma^{\Psi}(i - i_h)$$
$$\omega = \Omega^{\Psi}(\theta)(i - i_h) \qquad (4.14)$$
$$\beta = \beta^{\Psi}(\kappa)\Gamma^{\Psi}(i - i_h)$$

The Height Model: The expected pixel height μ (see Equation (4.13)) varies linearly with vertical image position i. Different height models $\Gamma^{\Psi}, \psi \in \Psi$ must be defined for each type of object ψ - see Figure 4-3(a). A further assumption is made that the projected height of vehicles does not depend on the orientation of the object.

The Width Model: For vehicular objects, the projected pixel width ω varies both as a function of depth (and hence varies linearly with position i) but also varies as a function of the 3D orientation of the object. The 3D orientation of a moving vehicle is correlated with the direction of its visual motion. This relationship can be clearly demonstrated in Figure 4-3(b) which plots 2D width (normalised by vertical height) against the visual motion direction θ for a large set of detected regions. Thus the projected width of an event is a function both i and the direction θ.

The Base Model: The vertical extent of the base again varies linearly with the vertical image position (Figure 4-3(c)). In dull weather conditions, this base area is usually a small fraction. However in bright weather conditions, this base area can be become significantly larger. Currently, the base model parameter β^{Ψ} is manually set as a proportion of the height model. Ideally some environmental illumination parameter κ would select the appropriate ratio.

4. REGISTERING MULTIPLE CAMERAS

In Section 2.1 the positions and velocity of objects tracked in each field of view were back-projected onto a local reference frame set on the ground plane. The transformation between cameras is unknown but it can be easily calculated if the correspondences between object positions are known between views [11]. In our auto-calibration scenario, we cannot assume that the correspondences are known. Further, while we assume the availability of object positions with associated velocity vectors, no temporal association is assumed.

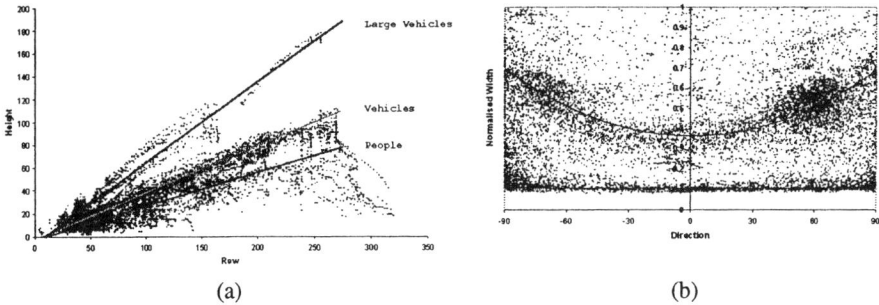

Figure 4-4. θ=0 indicates vertical motion, while θ=π/2 refers to horizontal motion. The ower plot illustrates that person width does not depend on orientation. For vehicles, the width increases from a minimum at θ=0 (face-on) to a maximum at θ=π/2 (side views). (a) Projected Height (b) Normalised Width.

The Hough Transform approach [1] has been adopted to recover the inter-camera transformation by taking advantage of the fact that the ground plane coordinate systems of temporally synchronised observations of the same 3D object are related by a simple rotation ψ and translation T transformation.

$$X' = R(\psi)X + T$$
$$V' = R(\psi)V \qquad (4.15)$$

where X, V and X', V' are positional and velocity observations measured in the local GPCS of two cameras C and C' respectively. Note that the velocity estimates are computed from the partial derivatives of Equation (4.2) respect to image coordinates, and the 2D tracker image position (i, j) and visual velocity (u, v) estimates *i.e.*

$$V_X = \frac{\partial X}{\partial i}u + \frac{\partial X}{\partial j}v, \quad V_Y = \frac{\partial Y}{\partial i}u + \frac{\partial Y}{\partial j}v \qquad (4.16)$$

In every frame interval, each camera outputs a set of measurements about all objects located in its field of view. As object correspondences are unknown, every pair of observations from each of the cameras must be used to generate a candidate estimate of the transformation. Given a pair observations $X'_{t,i}, V'_{t,i}$ and $X_{t,j}, V_{t,j}$ at time t from camera C' and C respectively, transformation estimates may be defined as

$$\cos\psi_{t,i,j} = \hat{V}'_{t,i} \cdot \hat{V}_{t,j}$$
$$T_{t,i,j} = X'_{t,i} - R(\psi_{t,i,j})X_{t,j} \qquad (4.17)$$

where $\hat{V}$ is the unit vector in the direction of V. If Λ_t and Λ'_t are the set of observations in frame t for cameras C and C' respectively, then the set of all observations

$$\{\psi_{\tau,i,j}, T_{\tau,i,j}; \forall i \in \Lambda'_\tau, \forall j \in \Lambda_\tau, \forall \tau \leq t\} \qquad (4.18)$$

should ideally exhibit a distinct cluster of estimates around the true solution ψ, T within an noise floor of uncorrelated false estimates generated by incorrectly corresponded observation pairs and noise observations. To detect this cluster, the space could be tessellated into bins and a Hough transform technique applied to locate the maximum that correspond to the optimal transformation parameters. However, the range of translation values required is difficult to predict a priori. Therefore to avoid the storage of the problems such a voting strategy introduces, a robust clustering approach is adopted. The expectation-maximisation mixture of Gaussian technique was implemented and adapted to iteratively perform the cluster analysis on the incoming stream of transform estimates. The clustering process continually reports the most likely transformations between cameras.

5. RESULTS

In the following sections we evaluate the three stages of the overall approach separately. In Section 5.1 the accuracy of the recovery of the local ground plane is tested by comparing the actual and estimated look-down

4. Learning the fusion of Video Data streams

angles. The Tsai calibration results performed on the PETS2001[2] were not particularly accurate at estimating the camera height and look-down angle. Consequently the evaluation was performed on the three local installations illustrated in Figure 4-5(a),(b) and (c).

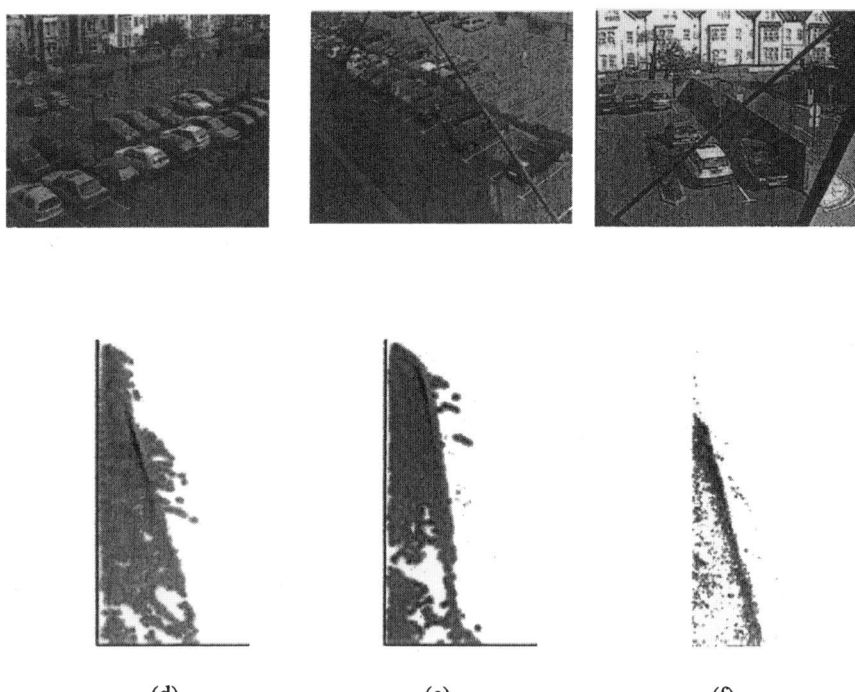

(d)   (e)   (f)

Figure 4-5. The TEST Installation Figures (d), (e) and (f) show the (inverted) histograms for the viewpoints illustrated in Figures (a) Camera 1, (b) Camera 2, and (c) Camera 3.

In Section 5.1 the *Ground Plane Tracker* (GPT) is compared to the *Image Plane Tracker* (IPT) and further summarised. Section 5.3 illustrates the process of camera ground plane registration, and evaluates the accuracy of the camera registration results on these and the PETS datasets.

5.1 Image to Ground Plane Calibration

The test installations illustrated in Figure 4.5 involve different types of camera placed at different heights overlooking a common car park scene. The car park has been surveyed to generated real-world ground plane

---

[2] The PETS2001 datasets (visualsurveillance.org) are problematic as they contain so few tightly distributed calibration points.

positions in a common coordinate system. These points have been selected to ensure that each camera has ten well distributed points in the image plane. The convex hull of these points contains most of the car park and over fifty percent of the visual plane. The real lookdown angles and camera heights have been established using surveying equipment from the ground plane projection of the correct optical axes.

As described in Section 2.3 the projected height model for each camera can be recovered by accumulating in a height versus vertical image position space and fitting a straight line to the resultant histogram. Results for Cameras 1, 2 and 3 are shown in Figures 4-5 (d), (e) and (f). In conjunction with the measured height of the cameras above the ground plane, the parameters of these models can be used to derive the extrinsic and some of the intrinsic camera parameters - see Equation (4.13). To compare the accuracy of the proposed method, the ground truth data, the traditional Tsai [11] technique results and the measurements generated by our approach are tabulated in Table 4-1. In all cases, the accuracy of the Tsai method and our own is comparable, with the shallow angle of view of Scene 3 being the most problematic. We employed the Tsai results to confirm that the camera had no significant roll *i.e.* rotation around the optical axis - typically less than 4°. The method proposed in this work accurately located the lookdown angle although care had to be taken to correctly fit the linear model to the projected height ridge of the histogram.

Table 4-1. Look-down Angle Results. For clarity the look-down angle has been redefined as defining the angle of intersection between ground plane and optical axis.

Test Installation	Camera 1	Camera 2	Camera 3
Correct Height	9.1m	13.9m	6.7m
Tsai Height	9.9m	15.4m	5.7m
HRE γ	0.195	0.109	0.255
Horizon i_h	-22.3	-174	17
Correct Angle	16°	24.3°	13.5°
Tsai Angle	16.7°	24.5°	7.7°
Our Approach	15.5°	23.3°	11.7°

5.2 Model Based Tracking

The *Ground Plane Tracker* (GPT) embeds the mechanisms introduced in this paper within a standard tracking framework, and is compared against a standard 2D tracker - the *Image Plane Tracker* (IPT). Both mechanisms employ a Kalman filter model whose observation and dynamic noise models

4. Learning the fusion of Video Data streams

are learnt directly from the data. The two methods are summarized in Table 4.2 below.

Table 4-2. Implementation details of standard and proposed tracking algorithms

Algorithm	Image Plane Tracker	Ground Plane Tracker
Measurement	x, y image pixels	X, Y ground plane-eq 1.2
Motion Model	First-order $x, y, \dot{x}, \dot{y}$	Second-order $X, Y, \dot{X}, \dot{Y}, \ddot{X}, \ddot{Y}$
Appearance Model	First-order Kalman filter on bounding box dimensions $h, w, \dot{h}, \dot{w}$	Position and velocity constrained bounding box model of Equation (4.13)

Data association is performed by searching predicted bounding boxes for union of overlapping moving regions whose area is greater than 10% of bounding box area. Model instances are instantiated from unassigned *moving regions* [10] whose areas are greater than some common threshold - 10 pixels (in quarter-size PAL frames). In neither case is any additional appearance matching implemented to improve data association. Observation position error is defined as deviation from predicted object dimension. Each object has a *time-to-live* counter (TTL) defined as $\min(TTL, 10)$ which is incremented if inter-frame match recovered, and decremented if no match recovered with object deleted when $TTL < 0$.

To compare the different approaches a *tracking error* is defined as the number of *track failures* per 1000 *track frames*. A track failure occurs when the tracking identity of any ground truth object changes. Track frames are the total number of object appearances for all tracks in a sequence. The experiment is run on three different datasets - see Figure 4-5: the PETS 2001 Dataset 1 (an occlusion rich dataset of distant objects in good lighting conditions), the *Kingston Car Park Dataset* (although relatively free of non-static occlusions, objects exhibit considerable motion variation against background undergoing frequent and severe lighting variations caused by intermittent direct and reflected sunshine), and the Football Dataset (large number of objects undergoing correlated and rapidly changing

Table 4-3. Tracking error.

Tracker	PETS	DIRC	Football
IPT	3.2	1.5	49
GPT	1.9	1.1	11

Note that the tracking results reflect the challenging nature of the Kingston datasets and, in particular, the Football Dataset. Nonetheless, the proposed tracker outperforms the traditional tracker which is easily misled. Greater insight into the problems of trackers can be gained by determining the nature and frequency[3] (% of frames) of the failure modes – see Table 4.4.

Table 4-4. Tracking error.

Failure	Description of data association failure	Failure IPT	GPT
Fragmentation	Unexpected small displaced observation	9%	2%
Static Occlusion	Unexpected small displaced observation	23%	10%
Object occlusion	Unexpectedly large observation	36%	34%
Motion model	Motion model too constraining	21%	34%
Stationary Object	Object merges into background	11%	20%

Both trackers loose track of objects that are stationary for several seconds determined by a TTL parameter.

However the principal weakness of the traditional tracker is when dealing with situations where (i) fragmentation or static occlusion processes shrink the search window with consequent failure to locate validating observations, and (ii) occlusions which widen the search window causing the tracker to be deflected by the occluding object. These problems are more likely in situations where the trajectory deviates from the assumed motion model.

5.3 Multi-Camera Calibration

Figure 4.6 plots the tracked object trajectories recovered from our motion detection and tracking software [8] and projected onto the local ground plane of each camera.

4. Learning the fusion of Video Data streams

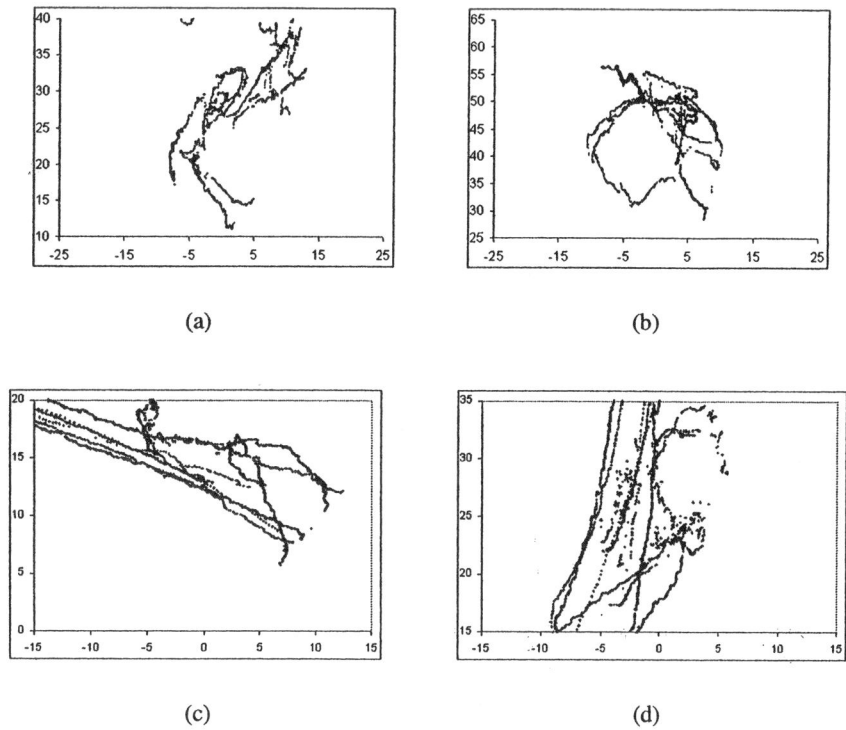

Figure 4-6. Projected trajectories: note only a roughly contemporaneous set of trajectories are plotted. (a) Test camera 1. (b) Test camera 2. (c) PETS camera 1. (d) PETS camera 2.

These observations are used to build the rotation and translation Hough space described by Equations (4.17). The populated space and dominant cluster are shown in Figure 4-7 (a) and (b) for the TEST and PETS datasets respectively. Note that these peaks are robustly recovered from an extensive noise floor. Produced by computing registration estimates for every pair of trajectory observations, this floor arises from the need to avoid the prior establishment of observation correspondences.

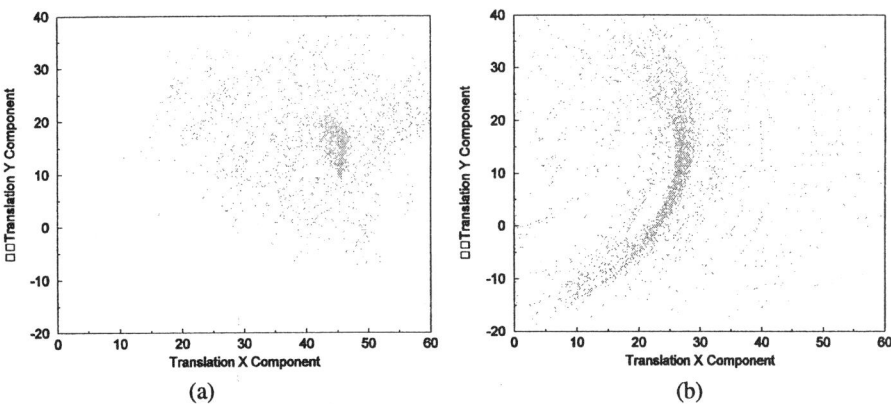

Figure 4-7. Locating the Maximal Cluster in Transform Space. (a) TEST. (b) PETS

The accuracy of the technique may be judged as before by comparing the registration results of the method with the equivalent data supplied by the Tsai calibration method (and the ground truth for the TEST datasets). Table 4.5 plots the rotation angle ψ (in degrees) and distance $|T|$ (in metres) between the origins of the two local GPCS for the TEST and PETS datasets.

Table 4-5. Registration results.

Datasets	Measurements			
	Tsai		Proposed	
	ψ	$|T|$	ψ	$|T|$
TEST	75°	57.2	81°	53.5
PETS	76°	27.9	70°	29.5

Despite the poor accuracy associated with off-ground-plane estimates, the accuracy of the ground plane projections are in agreement with the correct values surveyed in the DIRC datasets. Thus only the Tsai results are quoted in Table 4-5. The recovered values for the rotation angle ψ and distance $|T|$ are used to rotate and translate the data into a single coordinate system (that of the second camera). The overlapped trajectories are displayed in Figure 4-8.

4. Learning the fusion of Video Data streams

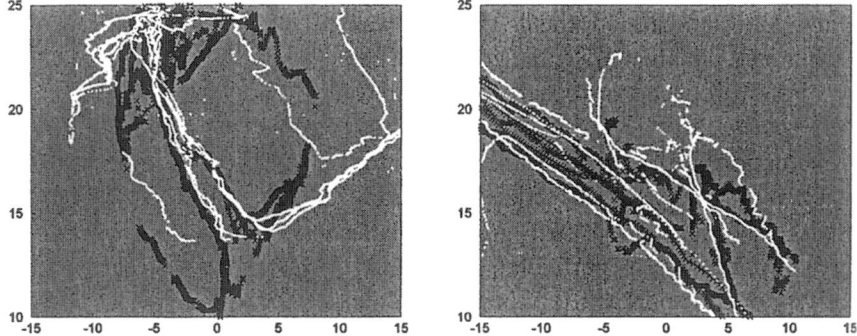

Figure 4-8. Overlaying Trajectories.

While not in perfect alignment, the accuracy appears sufficient to establish the correspondence of any new objects that enter the scene. Any lack of alignment arises from a number of sources: (i) any existing roll angle on either camera; (ii) inaccuracies in estimation of the lookdown angle and intrinsic parameters of either camera; and (iii) view-dependent positional bias in trajectories. The presented results are based on the location of the foot of an object on the ground plane. This position has demonstrated a strong viewpoint dependency when applied to car objects or person objects in the presence of shadows. A more consistent vertically weight centroid position will increase the degree of alignment.

6. CONCLUSIONS

A central objective of this work focuses on the development of learning techniques for use in plug-and-play visual surveillance multi-camera systems. Many camera calibration techniques exist, however most of them require the assistance of an expert to tune a set of parameters. The underlying strategy is to develop a suite of algorithms that could be installed by non-technical personnel, and as much as possible based on self-adjusting techniques that learn how to adapt to the camera set-up, the environmental changes and possibly to weather conditions. This paper proposes part of this work. This paper presents a novel camera calibration approach, based on two separate stages.

In the first stage a linear model of the projected height of objects in the scene is used in conjunction with world knowledge about the average person height and the height of each camera to recover the image-plane to local-ground-plane transformation of each camera. In the second stage, a clustering technique (based on expectation-maximisation) is then used to recover the transformation between these local ground planes. A comparison between the proposed technique and the standard approach of Tsai was carried out. Results for both techniques, evaluated with ground truth measures, show that the accuracy of the proposed approach is similar to Tsai's approach.

Although a more detailed evaluation is required, the presented preliminarily results demonstrate that approach generates sufficient accuracy to enable trajectory data to be fused within a common ground plane coordinate system between each pair of cameras. In particular, to robustly support the *plug and play* the sensitivity of the approach to violations in the assumptions of (i) projected height linearity and (ii) zero-roll angle must be investigated. Finally, the method had to be tested on a new data set rather than the PETS2001 images as the lack of calibration points makes the recover of accurate camera height problematic.

ACKNOWLEDGEMENT

The work was funded by the British Council and MIUR/CRUI organisations under the British-Italian Partnership Programme.

REFERENCES

[1] Dana H. Ballard and Christopher M. Brown. *"Computer Vision"*. Prentice-Hall, Inc., New Jersey, 1982.

[2] Y. Bar-Shalom and T. Fortmann. *"Tracking and Data Association"*. Mathematics in Science and Engineering. Academic Press, 1988.

[3] F. Cupillard, F. Bremond, and M. Thonnat. "Tracking Groups of People For Video Surveillance". In *2nd European Workshop on Advanced Videobased Surveillance Systems*, pages 88–100, Kingston, UK, September 4, 2001.

[4] T. Ellis and M. Xu. "Object Detection and Tracking in an Open and Dynamic World". In *Second IEEE International Workshop on Performance Evaluation of Tracking and Surveillance*, Hawaii, December 2001.

[5] L.M. Fuentes and S.A. Velastin. "People Tracking in Surveillance Applications".In *Second IEEE International Workshop on Performance Evaluation of Tracking and Surveillance*, Hawaii, December 2001.

[6] R.M. Haralick and H. Joo. "2D-3D Pose Estimation". *Proceedings of the International Conference on Pattern Recognition*, pages 385–391, November 14-17 1988.

[7] I.Haritaoglu, D.Harwood, and L.S.Davis. "W4: Real-time Surveillance of people and their Activities". *IEEE Transactions on Pattern Analysis and Machine Intelligence*, 19(7):809–830, July 1997.

[8] J. Orwell, P. Remagnino, and G.A. Jones. "From Connected Components to Object Sequences". In *First IEEE International Workshop on Performance Evaluation of Tracking and Surveillance*, pages 72–79, 2000.

[9] N.T. Siebel and S.J. Maybank. "Real-time Tracking of Pedestrians and Vehicles". In *Second IEEE International Workshop on Performance Evaluation of Tracking and Surveillance*, Hawaii, December 2001.

[10] C. Stauffer and W.E.L. Grimson. "Learning Patterns of Activity using Real-Time Tracking". *IEEE Transactions on Pattern Analysis and Machine Intelligence*, 22(8):747–757, August 2000.

[11] Roger Y. Tsai. "A versatile Camera Calibration Technique for High- Accuracy 3D Machine Vision Metrology Using Off-the-Shelf TV Cameras and Lenses". *IEEE Journal of Robotics and Automation*, RA-3(4):323–344, August 1987.

[12] T-H. Chang and S. Gong, "Bayesian Modality Fusion for Tracking Multiple People with a Multi-Camera System", Chapter 6, Video-Based Surveillance Systems: Computer Vision and Distributed Processing, pp. 79-87, Kluwer Publisher, 2002.

[13] J.Black, T.Ellis and P.Rosin, "Multi-View Image Surveillance and Tracking", to appear in IEEE Workshop on Motion and Video Computing, Orlando, Dec. 2002.

[14] O.Faugeras, "Three-Dimensional Computer Vision: a Geometric Viewpoint", MIT Press, 1993.

[15] R.Hartley and A.Zisserman, "Multiple View Geometry in Computer Vision", Cambridge Press, 2001.

Chapter 5

IMAGE FUSION USING THE EXPECTATION-MAXIMIZATION ALGORITHM AND A GAUSSIAN MIXTURE MODEL

Rick S. Blum and Jinzhong Yang
Lehigh University

Key words: Image fusion, Gaussian model, Concealed weapon detection, Autonomous landing guidance

1. INTRODUCTION

Image fusion refers the process of combining multiple images of a scene to obtain a single composite image [1-4]. The different images to be fused can come from different sensors of the same basic type or they may come from different types of sensors. The composite image should contain a more useful description of the scene than provided by any of the individual source images. This fused image should be more useful for human visual or machine perception. In recent years, image fusion has become an important and useful technique for image analysis, computer vision [4-7], concealed weapon detection (CWD) [8,9], and autonomous landing guidance (ALG) [10-11].

A simple image fusion method is to take the average of the source images pixel by pixel. While simple this approach can produce several undesired side effects including reduced contrast. In recent years, many researchers recognized that multiscale transforms are very useful for analyzing the information content of images for the purpose of fusion [12-14]. Several multiscale transforms have become very popular. These include the Laplacian pyramid transform [15], the contrast pyramid transform [16-17], the gradient pyramid transform [18], and the wavelet transform [12,14,19]. At the same time, some sophisticated image fusion approaches based on multiscale representations began to emerge and receive increased attention [12-19]. Figure 5-1 illustrates the block diagram of a generic image fusion scheme based on using multiscale decomposition (MSD) [14]. The basic idea is to perform a multiscale transform (MST) on each source image, then construct a composite multiscale representation from these according to some specific fusion rules. The fused image is obtained by taking an inverse multiscale transform (IMST).

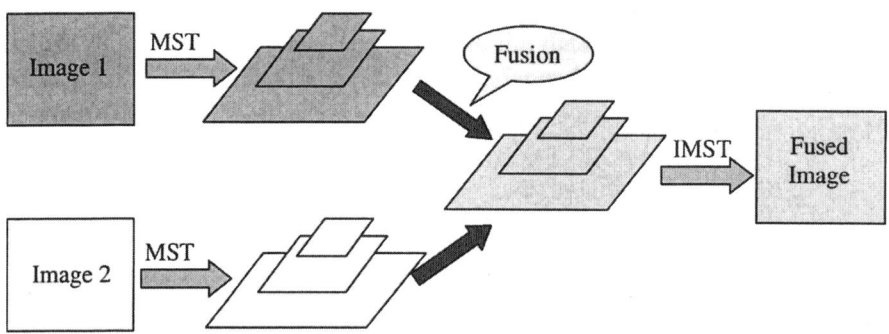

Figure 5-1. Block diagram of a generic image fusion scheme.

The majority of current image fusion methods that have been proposed have not been developed using a rigorous application of estimation theory. One exception is the work of Sharma and his co-authors [20-21] who has proposed a Bayesian fusion method which is based on estimation theory and assumes all distortions follow a Gaussian distribution. Since this is a rather limiting assumption, here we present a generalization that allows both Gaussian and non-Gaussian distortion based on which assumption best fits the observed data. The fusion takes place in the MST domain and the Laplacian pyramid transform is used [15].

The sensor images are modeled as the true scene corrupted by Gaussian mixture distortions as described in Section 2. As discussed in Section 3, we use the expectation-maximization (EM) algorithm to develop iterative equations to estimate the model parameters and the fused result. Section 4 presents experimental results and conclusions are given in Section 5.

2. IMAGE FORMATION MODEL

We assume that there exists a true scene which gives rise to a sensor image through a mapping. This mapping can be approximated by a locally affine transformation. This transformation is defined at every coefficient of the MST as

5. Image fusion using the expectation-maximization algorithm and a Gaussian mixture

$$z_i(j) = \beta_i(j)s(j) + \varepsilon_i(j) \qquad i = 1,\ldots,q \qquad (5.1)$$

where $i=1,\ldots,q$ indexes the sensors, j denotes the coefficient location (for example, it is shorthand for $j \equiv (x, y, m)$ where x, y are the pixel coordinates, and m is the level of the pyramid), $z_i(j)$ is the observed sensor image, $s(j)$ is the true scene (which we hope to approximate using fusion), $\beta_i(j) = \pm 1$ or 0 is the sensor selectivity factor, and $\varepsilon_i(j)$ is the random distortion. This model acknowledges that a given sensor may be able to "see" certain objects ($\beta_i(j) = 1$), may fail to "see" other objects ($\beta_i(j) = 0$), or may "see" certain objects with a polarity-reversed representation ($\beta_i(j) = -1$). The distortion is modeled as a K-term mixture of Gaussian probability density functions (pdfs) as

$$f_{\varepsilon_i(j)}(\varepsilon_i(j)) = \sum_{k=1}^{K} \lambda_{k,i}(j) \frac{1}{\sqrt{2\pi\sigma_{k,i}^2(j)}} \exp\left(-\frac{\varepsilon_i(j)^2}{2\sigma_{k,i}^2(j)}\right) \qquad (5.2)$$

This image formation model is generally different for each coefficient j. However, to a first order approximation the selectivity factor and the parameters of the pdf in (5.2) can be considered constant over a small region of neighboring coefficients, a fact that will be used during the parameter estimation.

3. IMAGE FUSION USING AN ITERATIVE EM-BASED ALGORITHM

The image formation model in (5.1) and (5.2) has been used in conjunction with the expectation-maximization (EM) algorithm [22-26] to develop a set of iterative equations to estimate the model parameters and to produce the fused image (using the final scene estimate). Since the EM algorithm is used, approximate maximum likelihood estimates [23] are produced. The iterative algorithm will be run once for each coefficient j to obtain the estimates of $s(j)$, $\beta_i(j)$, and $\{\lambda_{1,i}(j),\ldots,\lambda_{K,i}(j); \sigma_{1,i}^2(j),\ldots,\sigma_{K,i}^2(j)\}$ for each i.

3.1 Local Analysis Window

Consider the estimation for the quantities associated with coefficient j. In order to improve estimate reliably, we incorporate the neighboring coefficients. We define a local analysis window R_L centered on coefficient

j as shown in Figure 5-2. The window R_L has a size $L = h \times h$. In performing the estimates for the quantities associated with coefficient j, the calculations will employ the coefficients $l = 1,...,L$ in the window centered on coefficient j. Using a first order approximation, the parameters $\beta_i(j)$ and $\{\lambda_{1,i}(l),...,\lambda_{K,i}(l); \sigma^2_{1,i}(l),...,\sigma^2_{K,i}(l)\}$ will be considered to be the same for each coefficient $l = 1,...,L$ in the local analysis window. Thus we drop the indices on these parameters, using $\beta_i(l) = \beta_i$ for example. When the final estimates associated with coefficient j are produced using the iterative algorithm, the estimates produced for β_i and $\{\lambda_{1,i},...,\lambda_{K,i}; \sigma^2_{1,i},...,\sigma^2_{K,i}\}$ for $i = 1,...,q$ will be assigned to the corresponding parameters for coefficient j from (5.1), the final estimates of the scene for all coefficients neighboring coefficients j are discarded, and the final estimate of the scene for coefficient j will be assigned to $s(j)$ from (5.1). The window size L must be carefully chosen to be large enough to allow good estimation but small enough to allow our assumptions to be reasonable.

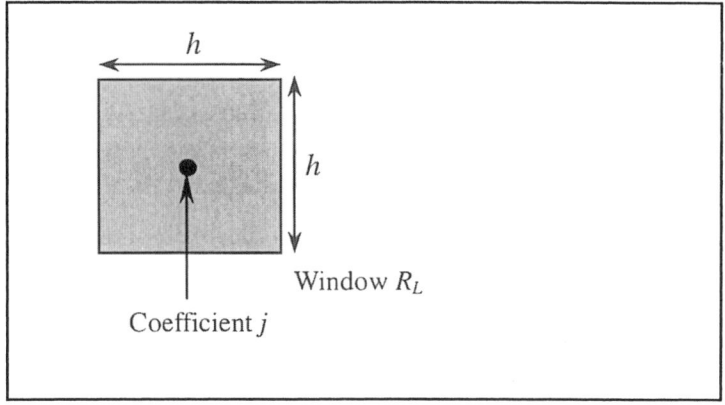

Figure 5-2. Using a window of data around coefficient j.

3.2 Iterative Fusion Procedure

We have developed a set of iterative equations using the well-developed theory of the EM algorithm. Let $s'(l)$ denote the updated value of $s(l)$, and assume similar notation for the other quantities β'_i, $\lambda'_{k,i}$ and $\sigma'^2_{k,i}$. The updated estimates are chosen to be those that maximize a likelihood-like function (see Appendix A for the details). The algorithm begins with the current estimates $s(l)$, β_i, $\lambda_{k,i}$, $\sigma^2_{k,i}$ of the parameters and produces updated estimates $s'(l)$, β'_i, $\lambda'_{k,i}$, $\sigma'^2_{k,i}$ at each step using the following procedure

5. Image fusion using the expectation-maximization algorithm and a Gaussian mixture

1. First compute $g_{k,il}[z_i(l)]$:

$$g_{k,il}[z_i(l)] = \frac{\dfrac{\lambda_{k,i}}{\sqrt{2\pi\sigma_{k,i}^2}}\exp\left(-\dfrac{(z_i(l)-\beta_i s(l))^2}{2\sigma_{k,i}^2}\right)}{\sum_{p=1}^{K}\dfrac{\lambda_{p,i}}{\sqrt{2\pi\sigma_{p,i}^2}}\exp\left(-\dfrac{(z_i(l)-\beta_i s(l))^2}{2\sigma_{p,i}^2}\right)} \quad (5.3)$$

$$k=1,...,K; i=1,...,q; l=1,...,L$$

2. Update the parameter β_i for $i=1,...,q$. β'_i is chosen to have the value from the set $\{0,-1,+1\}$ that maximizes

$$Q = -\frac{1}{2}\sum_{i=1}^{q}\sum_{l=1}^{L}\sum_{k=1}^{K}\left[\ln(\sigma_{k,i}^2) + \frac{(z_i(l)-\beta'_i s(l))^2}{\sigma_{k,i}^2}\right]\cdot g_{k,il}[z_i(l)] \quad (5.4)$$

3. Recalculate $g_{k,il}[z_i(l)]$ using (5.3) with the updated value β'_i. Then update the true scene $s(l)$:

$$s'(l) = \frac{\sum_{i=1}^{q}\sum_{k=1}^{K} z_i(l)\beta'_i \dfrac{g_{k,il}[z_i(l)]}{\sigma_{k,i}^2}}{\sum_{i=1}^{q}\sum_{k=1}^{K}\beta'^2_i \dfrac{g_{k,il}[z_i(l)]}{\sigma_{k,i}^2}} \qquad l=1,...,L \quad (5.5)$$

4. Recalculate $g_{k,il}[z_i(l)]$ using (5.3) with the updated value β'_i and $s'(l)$. Then update the distortion parameters $\lambda_{k,i}$ and $\sigma_{k,i}^2$:

$$\lambda'_{k,i} = \frac{1}{L}\sum_{l=1}^{L} g_{k,il}[z_i(l)] \qquad k=1,...,K; i=1,...,q \quad (5.6)$$

$$\sigma'^2_{k,i} = \frac{\sum_{l=1}^{L}(z_i(l)-\beta'_i s'(l))^2 g_{k,il}[z_i(l)]}{\sum_{l=1}^{L} g_{k,il}[z_i(l)]} \qquad k=1,...,K; i=1,...,q \quad (5.7)$$

5. Repeat steps 1 to 4 using the new parameters $s'(l)$, β'_i, $\lambda'_{k,i}$ and $\sigma'^2_{k,i}$.

The above steps 1-5 are derived from the SAGE version of the EM algorithm [24], similar to the development in [25]. The details of the derivation are presented in Appendix A. Note that the updates for $\lambda_{k,i}$ in (5.6) implicitly contain the constraint $\sum_{k=1}^{K}\lambda_{k,i}=1$, for $i=1,...,q$. Further, it is clear from the sums over the variable l that (5.4), (5.6) and (5.7) perform calculations over the analysis window.

The above algorithm is capable of modeling both Gaussian and non-Gaussian distortion depending on which better models the observed images using (5.1) and (5.2). If a Gaussian distortion model is employed only one of the terms in the sum $\sum_{k=1}^{K}\lambda_{k,i}=1$ will be 1 and the other terms will be zero. This leads to $g_{k,il}[z_i(l)]=1$ in (5.3) and the estimates in (5.5) and (5.7) are exactly the standard Gaussian maximum likelihood estimates. For example in (5.5) we estimate the scene by a weighted average of the observed sensor images which contain the scene (based on β'_i being nonzero). A sensor image is weighted by a lower weight if the variance of its corresponding distortion is larger as would be expected. When a non-Gaussian distortion model is more suitable, the function $g_{k,il}[z_i(l)]$ acts as a classifier which tries to estimate the probability $\lambda'_{k,i}$ that coefficient j has distortion that is best modeled as coming from the kth term in the Gaussian mixture model in (5.2) which explains the form of (5.6). Thus generally the iterative equations are attempting to determine which term in the mixture model best represents each piece of observed data and also which set of Gaussian functions are best for representing all of the observed data. Since the term classification has come up it is not surprising that (5.5)-(5.7) are equivalent to using a radial-basis function neural network [27], but (5.5)-(5.7) produce approximate maximum likelihood estimates. Thus we have an approach that gives the universal approximation capabilities and efficiency of a neural network approach but we also can interpret the fused result as optimizing a well-accepted estimation criterion.

3.3 Initialization of the Fusion Algorithm

Initial estimates are required for (5.3)-(5.7). We choose the initial estimates for the true scene $s(l)$ to come from the weighted average of the sensor images as per

$$s(l) = \sum_{i=1}^{q} w_i(l) z_i(l) \qquad l=1,...,L \qquad (5.8)$$

5. Image fusion using the expectation-maximization algorithm and a Gaussian mixture

where $\sum_{i=1}^{q} w_i(l) = 1$. In order to determine the $w_i(l)$ in (5.8), we employ a salience measure that was discussed in [28]. For each coefficient $l \equiv (x, y, m)$, the salience measure is

$$\Omega_i(l) = \sum_{x'=-2}^{2}\sum_{y'=-2}^{2} p(x', y') z_i^2(x+x', y+y', m) \quad i=1,...,q \qquad (5.9)$$

where $p(x', y')$ is the weight for each coefficient around l. Here a 5 by 5 window of coefficients centered on l are used to calculate the salience measure of l using

$$\mathbf{p} = \begin{pmatrix} 1/48 & 1/48 & 1/48 & 1/48 & 1/48 \\ 1/48 & 1/24 & 1/24 & 1/24 & 1/48 \\ 1/48 & 1/24 & 1/3 & 1/24 & 1/48 \\ 1/48 & 1/24 & 1/24 & 1/24 & 1/48 \\ 1/48 & 1/48 & 1/48 & 1/48 & 1/48 \end{pmatrix} \qquad (5.10)$$

It is reasonable to specify a larger weight for the central coefficients and a smaller weight for the outer coefficients. Then the $w_i(l)$, for $i=1,...,q$, are specified by the salience measure using

$$w_i(l) = \Omega_i(l) / \sum_{j=1}^{q} \Omega_j(l) \quad i=1,...,q \qquad (5.11)$$

A simple initialization for β_i is to assume that the true scene appears in each sensor image. Hence $\beta_i = 1$ for $i=1,...,q$. In order to model the distortion in a robust way the distortion is initialized as impulsive [25]. We initialize the distortion parameters with $\lambda_{1,i} = 0.8$ and $\lambda_{2,i} = \cdots = \lambda_{K,i} = 0.2/(K-1)$ for $i=1,...,q$. Then we set $\sigma_{k,i}^2 = \gamma \sigma_{k-1,i}^2$ for $i=1,...,q$, $k=2,...,K$, where the value for $\sigma_{1,i}^2$ is chosen based on an estimate of the total variance $\sigma_i^2 = \sum_{k=1}^{K} \lambda_{k,i} \sigma_{k,i}^2$ for $i=1,...,q$ given by

$$\sigma_i^2 = \sum_{l=1}^{L} [z_i(l) - s(l)]^2 / L \qquad (5.12)$$

We choose $\gamma = 10$ so that the initial distortion model is fairly impulsive. This initialization scheme worked very well for the cases we have studied. We observed that the algorithm in our experiments generally converges in 3 to 5 iterations.

3.4 Consistency Verification

Consider the jth coefficient in the pyramid transform. The estimates of β for this coefficient for each sensor are collected in the vector $\boldsymbol{\beta}_j = (\beta_{j1},...,\beta_{jq})$. Suppose $\boldsymbol{\beta}_l = (\beta_{l1},...,\beta_{lq}), l=1,...,L$ are the corresponding vectors for those coefficients around j. These L coefficients are those in a $h \times h$ local window centered on coefficient j. The consistency verification attempts to make sure that the β values for coefficient j are as close as possible to those of its neighbors. Thus the consistency verification selects the $\boldsymbol{\beta}_j$ minimizing

$$\Delta = \sum_{l=1}^{L} \|\boldsymbol{\beta}_j - \boldsymbol{\beta}_l\| \qquad (5.13)$$

In fact this minimization can be simplified by performing the calculation on a component-by-component basis.

4. EXPERIMENTAL RESULTS

We have applied our new fusion algorithm to a CWD application. CWD is an increasingly important topic in the general area of law enforcement and image fusion has been identified as a key technology to enable progress on this topic [8,9]. With the increasing threat of terrorism, CWD is a very important technology.

4.1 CWD with Visual and MMW images

The first example for a CWD application is to fuse the visual and MMW images shown in Figure 5-3 (a) and (b). The true scene and the parameters are initialized as described in Section 3.3. The fusion algorithm used the settings shown in Table 5-1.

Table 5-1. EM fusion settings for fusing visual and MMW images.

Sensor Images	2
Image size	256×256 pixels
Number of Laplacian pyramid levels	5
Gaussian mixture terms	2
Local window size	3×3 pixels

Figure 5-3 (c) shows the result of the EM fusion algorithm. Figure 5-3 (d), (e) and (f) show the results obtained respectively by pixel averaging, selecting the maximum pixel, and the using the Laplacian pyramid fusion

5. Image fusion using the expectation-maximization algorithm and a Gaussian mixture

approach in [14]. The fusion method from [14] uses 5 Laplacian pyramid levels and chooses the maximum sensor pyramid coefficients for the high-pass pyramid coefficients and averages the sensor pyramid coefficients for the low-pass pyramid coefficients. From the comparison, the EM fusion algorithm performs better than the other 3 fusion methods. From the fused image, there is considerable evidence to suspect that the person on the right has a concealed gun beneath his clothes. This fused image may be very helpful to a police officer, for example, who must respond promptly.

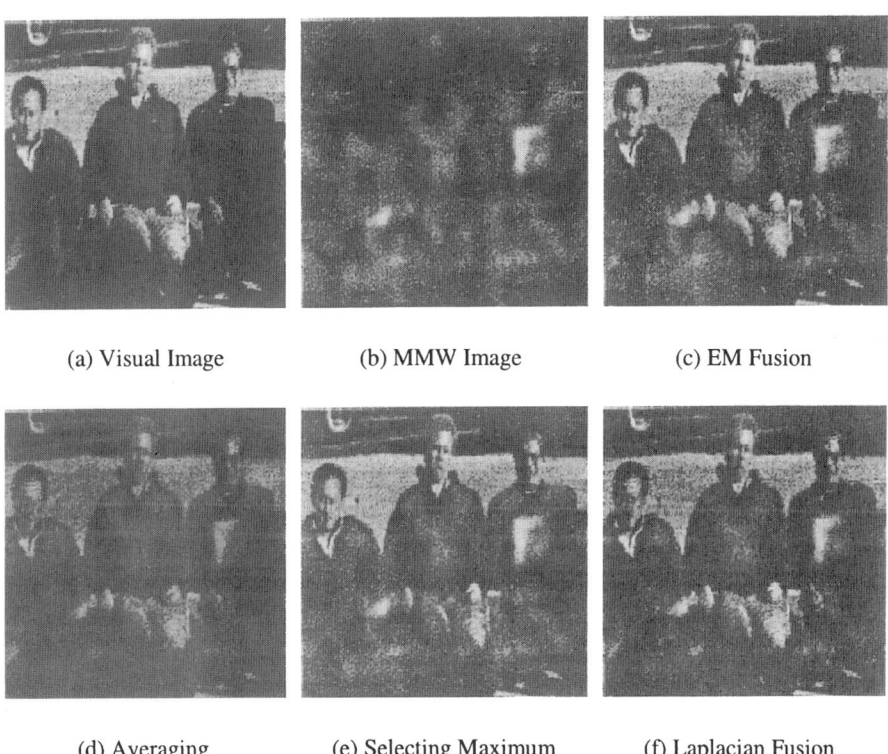

(a) Visual Image   (b) MMW Image   (c) EM Fusion

(d) Averaging   (e) Selecting Maximum   (f) Laplacian Fusion

Figure 5-3. Visual and MMW images and fused result for CWD.

4.2 CWD with Visual and IR images

Table 5-2. EM fusion settings for fusing visual and IR images.

Sensor Images	2
Image size	256×256 pixels
Number of Laplacian pyramid levels	7
Gaussian mixture terms	2
Local window size	5×5 pixels

The second example considers a CWD application employing a visual and an IR image. Figure 5-4 (a) and (b) show the visual and IR images. Figure 5-4 (c), (d), (e) and (f) show the fused result obtained by using the EM fusion algorithm, pixel averaging, selecting the maximum pixel, and the algorithm from [14] respectively. In the EM fusion algorithm, the true scene and the parameters are initialized as described in Section 3.3. The settings used in the EM fusion implementation are shown in Table 5-2. This example shows that the EM fusion algorithm performs well for this example. From the comparison, the EM fusion algorithm performs better than pixel averaging, selecting the maximum pixel, and the algorithm from [14].

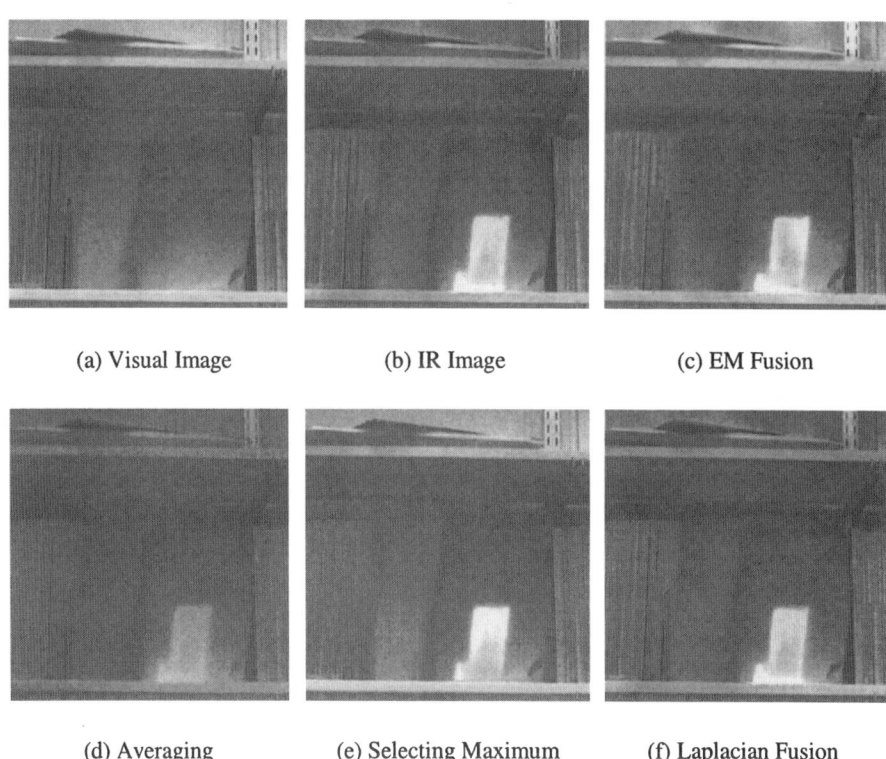

(a) Visual Image (b) IR Image (c) EM Fusion

(d) Averaging (e) Selecting Maximum (f) Laplacian Fusion

Figure 5-4. Visual and IR images and fused result for CWD.

4.3 Fusion for Autonomous Landing Guidance (ALG) applications

An autonomous landing guidance (ALG) system employs computer vision technology for landing an aircraft in bad weather. It also provides navigation guidance to pilots for landing the aircraft even in low visibility conditions [20]. In order to provide enhanced performance, image fusion is a

5. Image fusion using the expectation-maximization algorithm and a Gaussian mixture

key technology for ALG. Several different sensors can be used in ALG. Here we will use long wave and medium wave images in our example.

Figure 5-5 (a) and (b) show the long wave and medium wave images. Figure 5-5 (c), (d), (e) and (f) show the fused result obtained by the EM fusion algorithm, pixel averaging, selecting the maximum pixel, and the algorithm from [14] respectively. In the EM fusion algorithm, the initialization is as described in Section 3.3. The settings used in the EM fusion implementation are shown in Table 5-3. From the comparison, the EM fusion algorithm performs better than pixel averaging and selecting the maximum pixel, while it has similar performance to the algorithm from [14].

Table 5-3. EM fusion settings used in ALG case.

Sensor Images	2
Image size	233×233 pixels
Number of Laplacian pyramid levels	3
Gaussian mixture terms	2
Local window size	5×5 pixels

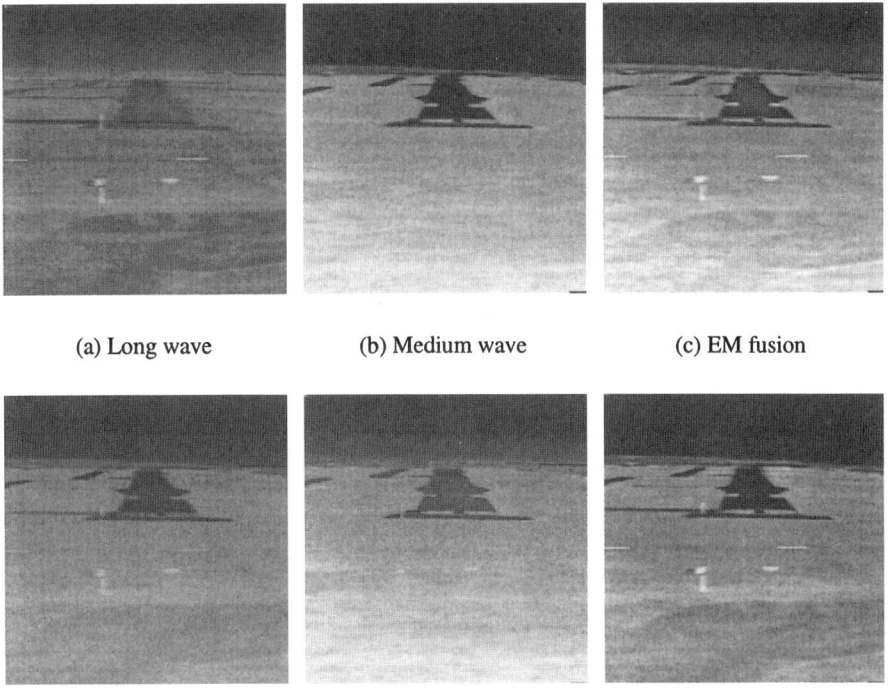

(a) Long wave   (b) Medium wave   (c) EM fusion

(d) Averaging   (e) Selecting Maximum   (f) Laplacian Fusion

Figure 5-5. Long wave and medium wave images and fused result for ALG.

5. CONCLUSION

We have presented a new image fusion method based on a Gaussian mixture distortion model and experimented with this method for concealed weapon detection applications and autonomous landing guidance applications. The results showed the advantages of this EM-based approach in some cases. We also have studied the effect of the settings (number of Laplacian pyramid levels, the size of local analysis window). Generally speaking, using more Laplacian pyramid levels can be beneficial but this comes at the cost of increased complexity. In practice, we have found there is not much difference between the fused results obtained using a different number of pyramid levels if a sufficient number of levels is used (5 in our experiments). The local analysis window should be small enough so that the parameters are indeed constant in the window, but it should be large enough to contain enough sensor data to estimate the parameters reliably. In cases we considered, we found a 5×5 or 3×3 window size is a good choice.

We have tested an extended version of our algorithm for cases where we have multiple sensor image frames taken from a video sequence and we have found that we can expect significant improvements in the fused result over the cases considered in this paper. The multiple frames will provide redundant information about the true scene. The additional information is very helpful in the estimation.

APPENDIX A: OUTLINE OF THE DERIVATION OF THE ITERATIVE ALGORITHM

The EM algorithm provides a general approach to iterative computation of maximum-likelihood estimates from incomplete data [22]. The first step in deriving the EM algorithm is the specification of a set of "complete data" and "incomplete data". For the image formation model (1) and (2), the incomplete data set $\mathbf{X}$ consists of the observed data as

$$\mathbf{X} = \{z_i(l) : i = 1,...,q; l = 1,...L\} \tag{5.14}$$

and the complete data set $\mathbf{X}_c$ is defined as

$$\mathbf{X}_c = \{(z_i(l), k_i(l)) : i = 1,...,q; l = 1,...L\} \tag{5.15}$$

where $k_i(l) \in \{1,2,...,K\}$ identifies which term in the Gaussian mixture pdf (5.2) produces the additive distortion sample in the observation $z_i(l)$. The common parameter set is $\Phi = \{s(l), \beta_i, \lambda_{k,i}, \sigma^2_{k,i}; l = 1,...,L; i = 1,...,q; k = 1,...,K\}$. The number of terms K in the Gaussian mixture pdf model in (5.2) is assumed to be fixed. Methods for choosing K have been previously considered [25].

5. Image fusion using the expectation-maximization algorithm and a Gaussian mixture

The SAGE version of the EM algorithm [24,25] is used to derive the iterative estimation equations. The elements of the incomplete data $\mathbf{X}$ in (14) are iid with marginal pdf

$$z_i(l)|\Phi \sim h_{il}(z_i(l)|\Phi) = \sum_{k=1}^{K} \frac{\lambda_{k,i}}{\sqrt{2\pi\sigma_{k,i}^2}} \exp\left(-\frac{(z_i(l)-\beta_i s(l))^2}{2\sigma_{k,i}^2}\right) \quad (5.16)$$

The elements of the complete data $\mathbf{X}_c$ in (5.15) are independent with marginal pdf

$$(z_i(l),k_i(l))|\Phi \sim h_{c,il}(z_i(l),k_i(l)|\Phi) = \frac{\lambda_{k_i(l)}}{\sqrt{2\pi\sigma_{k_i(l)}^2}} \exp\left(-\frac{(z_i(l)-\beta_i s(l))^2}{2\sigma_{k_i(l)}^2}\right) \quad (5.17)$$

and the conditional distribution $k_i(l)|z_i(l),\Phi$ is

$$g_{k,il}[z_i(l)] = \frac{h_{c,il}(z_i(l),k_i(l)|\Phi)}{h_{il}(z_i(l)|\Phi)} = \frac{\dfrac{\lambda_{k,i}}{\sqrt{2\pi\sigma_{k,i}^2}}\exp\left(-\dfrac{(z_i(l)-\beta_i s(l))^2}{2\sigma_{k,i}^2}\right)}{\sum_{p=1}^{K}\dfrac{\lambda_{p,i}}{\sqrt{2\pi\sigma_{p,i}^2}}\exp\left(-\dfrac{(z_i(l)-\beta_i s(l))^2}{2\sigma_{p,i}^2}\right)} \quad (5.18)$$

$$k=1,\ldots,K; i=1,\ldots,q; l=1,\ldots,L$$

The joint pdfs for the incomplete and complete data sets are $h(\mathbf{X}|\Phi) = \prod_{i=1}^{q}\prod_{l=1}^{L}h_{il}(z_i(l)|\Phi)$ and $h_c(\mathbf{X}_c|\Phi) = \prod_{i=1}^{q}\prod_{l=1}^{L}h_{c,il}(z_i(l),k_i(l)|\Phi)$, respectively.

Each iteration of the EM algorithm involves two steps, the expectation step (E-step) and the maximization step (M-step) [22] The E-step of the EM algorithm performs an average over the complete data, conditioned upon the incomplete data to produce the cost function

$$\begin{aligned}Q(\Phi'|\Phi) &= E\{\ln h_c(\mathbf{X}_c|\Phi')|\mathbf{X},\Phi\}\\ &= \sum_{i=1}^{q}\sum_{l=1}^{L}E\{\ln h_{c,il}(z_i(l),k_i(l)|\Phi')|z_i(l),\Phi\}\\ &= B + \sum_{i=1}^{q}\sum_{l=1}^{L}\sum_{k=1}^{K}\ln(\lambda'_{k,i})\cdot g_{k,il}[z_i(l)]\\ &\quad -\frac{1}{2}\sum_{i=1}^{q}\sum_{l=1}^{L}\sum_{k=1}^{K}\left[\ln(\sigma'^2_{k,i})+\frac{(z_i(l)-\beta'_i s'(l))^2}{\sigma'^2_{k,i}}\right]\cdot g_{k,il}[z_i(l)]\end{aligned} \quad (5.19)$$

where B is a term that is independent of Φ The EM algorithm would update the parameter estimates to new values Φ' that maximize $Q(\Phi'|\Phi)$ in (5.19). This is called the M-step of the EM algorithm.

In order to maximize $Q(\Phi'|\Phi)$ analytically, we update each parameter one at a time. Because β_i is discrete, β'_i is updated to have the value from the set $\{0, -1, +1\}$ that maximizes equation (19) with all the other parameters set at their old values: $s'(l) = s(l)$, $\lambda'_{k,i} = \lambda_{k,i}$ and $\sigma'^2_{k,i} = \sigma^2_{k,i}$. The updated estimate for true scene $s'(l)$ in (5.5) is obtained from maximizing (5.19) analytically by solving $\partial Q/\partial s(l) = 0$ using the updated β' and the other old parameter values. The updated estimate for $\lambda'_{k,i}$ and $\sigma'^2_{k,i}$ in (5.6) and (5.7) are obtained from solving $\partial Q / \partial \lambda_{k,i} = 0$ and $\partial Q / \partial \sigma^2_{k,i} = 0$ respectively for $k = 1,...,K$, $i = 1,...,q$ using the updated parameters where possible.

REFERENCES

[1] P. K. Varshney, "Scanning the issue: Special issue on data fusion," *Proc. IEEE*, vol. 85, no. 1, pp. 3-5, Jan. 1997.

[2] D. L. Hall, *Mathematical Techniques in Multisensor Data Fusion*. Artech House, Boston, MA, 1992.

[3] E. Waltz and J. Llinas, *Multisensor Data Fusion*. Artech House, Boston, MA, 1990.

[4] D. L. Hall and J. Llinas, "An introduction to multisensor data fusion," *Proc. IEEE*, vol. 85, no. 1, pp. 6-23, Jan. 1997.

[5] P. K. Varshney, "Multisensor data fusion," *Electronics & Communication Engineering Journal*, vol. 9, pp. 245-253, Dec. 1997.

[6] J. K. Aggarwal, *Multisensor Fusion for Computer Vision*. Springer Verlag, 1993.

[7] L. A. Klein, *Sensor and Data Fusion Concepts and Applications*. SPIE, 1993.

[8] D. D. Ferris Jr., R. W. McMillan, N. C. Currie, M. C. Wicks, and M. A. Slamani, "Sensors for military special operations and law enforcement applications," *Proc. SPIE*, vol. 3062, pp. 173-180, 1997.

[9] M. A. Slamani, L. Ramac, M. Uner, P. K. Varshney, D. D. Weiner, M. G. Alford, D. D. Ferris Jr., and V. C. Vannicola, "Enhancement and fusion of data for concealed weapons detection," *Proc. SPIE*, vol. 3068, pp. 8-19, 1997.

[10] M. R. Franklin, "Application of an autonomous landing guidance system for civil and military aircraft," *Proceedings of SPIE*, vol. 2463, pp. 146-153, 1995.

[11] J. R. Kerr, D. P. Pond, and S. Inman, "Infrared-optical multisensor for autonomous landing guidance," *Proceedings of SPIE*, vol. 2463, pp. 38-45, 1995.

[12] Z. Zhang, *Investigations of Image Fusion*. PHD dissertation, Lehigh University, Bethlehem, PA, May 1999.

[13] Z. Zhang and R. S. Blum, "A hybrid image registration technique for a digital camera image fusion application," *Information Fusion*, pp. 135-149, June 2001.

[14] Z. Zhang and R. S. Blum, "A categorization and study of multiscale-decomposition-based image fusion schemes," *Proceedings of the IEEE*, pp. 1315-1328, Aug. 1999.

5. Image fusion using the expectation-maximization algorithm and a Gaussian mixture

[15] P. J. Burt and E. Adelson, "The Laplacian pyramid as a compact image code," *IEEE Trans. Communications*, vol. 31, no. 4, pp. 532-540, Apr. 1983.

[16] A. Toet, "Image fusion by a ratio of low-pass pyramid," *Pattern Recognition Letters*, vol. 9, no. 4, pp. 245-253, 1989.

[17] A. Toet, L. J. van Ruyven, and J. M. Valeton, " Merging thermal and visual images by a contrast pyramid," *Optical Engineering*, vol. 28, no. 7, pp. 789-792, July 1989.

[18] P. J. Burt, "A gradient pyramid basis for pattern-selective image fusion," *Society for Information Display, Digest of Technical Papers*, pp. 467-470, 1992.

[19] S. G. Mallat, "Multifrequency channel decompositions of images and wavelet models," *IEEE Trans. on Acoustic Speech Signal Processing*, vol. 37, no. 12, pp. 2091-2110, Dec. 1989.

[20] R. K. Sharma, *Probabilistic Model-based Multisensor Image Fusion*. PHD dissertation, Oregon Graduate Institute, Portland, OR, October 1999.

[21] R. K. Sharma, T. K. Leen, M. Pavel, "Probabilistic Image Sensor Fusion", *Advances in Neural Information Processing Systems*, vol. 11, The MIT Press, 1999.

[22] A. P. Dempster, N. M. Laird, D. B. Rubin, "Maximum likelihood from incomplete data via the EM algorithm," *J. of the Royal Statistical Soc. B*, vol. 39, no. 1, pp. 1-38, 1977.

[23] R. A. Redner and H. F. Walker, "Mixture densities, maximum likelihood and the EM algorithm," *SIAM Review*, vol. 26, no. 2, pp. 195-239, April 1984.

[24] J. A. Fessler and A. O. Hero, "Space-alternating generalized expectation-maximization algorithm," *IEEE Trans. Signal Processing*, vol. 42, no. 10, pp. 2664-2677, Oct. 1994.

[25] R. S. Blum, R. J. Kozick, and B. M. Sadler, "An adaptive spatial diversity receiver for non-Gaussian interference and noise," *IEEE Trans. Signal Processing*, vol. 47, no. 8, pp. 2100-2111, Aug. 1999.

[26] R. S. Blum, R. J. Kozick, and B. M. Sadler, "EM-based approaches to adaptive signal detection on fading channels with impulsive noise," *31th Annual Conference on Information Sciences and Systems*, Johns Hopkins University, pp. 112-117, Baltimore, MD, March 1997.

[27] I. Cha and S. A. Kassam, "RBFN Restoration of Nonlinearly Degraded Images," *IEEE Trans. Image Processing*, vol. 5, no. 6, pp. 964-975, June 1996.

[28] P. J. Burt and R. J. Kolczynski, "Enhanced image capture through fusion," in *Proc. the 4th Intl. Conf. on Computer Vision*, pp. 173–182, May 1993.

Chapter 6

VIDEO-BASED SURVEILLANCE FOR CHEM-BIO PROTECTION OF BUILDINGS

Ioannis Pavlidis[1], Christos Stathopoulos[2], Tony Faltesek[3]

[1]*University of Houston, Dept. of Computer Science, Houston, TX 77204-3010*
[2]*University of Houston, Dept. of Biology and Biochemistry, Houston, TX 77204-5001*
[3]*Honeywell Laboratories, 3660 Technology Drive, Minneapolis, MN 55418*

Key words: Video surveillance, Chem-bio protection, Event detection.

1. INTRODUCTION

In the wake of recent global events the threat of a chemical and biological (chem-bio) attacks became very real. The community of technology and policy experts has recognized this threat way ahead of its time and a very extensive treatise of the matter can be found at [1]. There are various scenarios that the terrorists may follow to stage a chem-bio attack. One such scenario by testimony of one terrorist is to attack through the air-intakes of commercial or Government buildings. In July 2001, the terrorist Ahmed Ressam described in his court testimony how the training he and many others received in Al-Qaeda camps included attacking buildings with chemical agents ([2]). In particular, he detailed how he was trained to use cyanide and sulfuric acid to create deadly fumes and how to inject these fumes into building air-intakes. The targeting of air-intakes meant to maximize the number of deaths in the building without creating a risk for the attacker.

The possible attack scenarios to air-intakes are low-tech and invariably involve a human. In a typical scenario the terrorist approaches the building air-intakes and unleashes the lethal load in the immediate proximity. Since human motion and activity is involved, it is possible to design a video-based surveillance system to automatically detect and report a possible threat. The threat will be reported on the guard's console. Ideally, if the level of threat is high, the system should also shut down the HVAC air handler to minimize casualties (**HVAC** stands for **H**eating, **V**entilation, and **A**ir **C**onditioning system). The staging of a chem-bio attack in a building is anticipated to last at most a few minutes. Therefore, an automated video surveillance system can provide only short notice. Nevertheless, even this short notice may make the difference between life and death if the air-hander of the building is shut down on time and the agent is blocked.

In this chapter we first describe the paralyzing effects of certain chem-bio agents (Section 2). Then, we describe various attack scenarios that an air-intake protection system should be capable of addressing (Section 3). In Section 4 we propose a system architecture for the protection of air-intakes. In Section 5 we describe the hardware and software components that we have developed so far and their performance. In Section 6 we conclude the chapter by outlining the additional hardware and software components that we are planning to develop and incorporate to our current baseline system.

2. CHEM-BIO AGENTS

Although there are more than forty chem-bio agents that could be used as weapons [3][4], the most probable ones to be used in attacking building air-intakes are the following five: hydrogen cyanide, anthrax, botulinum toxin, plague, and smallpox. All five are highly toxic or infectious after inhalation and have been weaponized by nations or individuals in several reported cases.

Hydrogen Cyanide: It s an extremely flammable, colorless gas or liquid. It gives off toxic fumes in a fire and is highly explosive. Exposure irritates the eyes, the skin and the respiratory tract. Symptoms are burning and redness for the skin and eyes. Inhalation causes confusion, drowsiness and shortness of breath, leading to collapse. The substance can affect the central nervous system, resulting in impaired respiratory and circulatory functions. Exposure can be fatal. Recommended antidotes include exposure to fresh air in the case of inhalation and rinsing with plenty of water in the case of skin or eye exposure.

Anthrax: It is an acute infectious disease caused by the Gram-positive bacterium *Bacillus anthracis*. There are several forms of human anthrax, but the serious ones are inhalation anthrax, cutaneous anthrax, and intestinal anthrax. Symptoms of the disease usually occur within seven days after infection. Initial symptoms of inhalation anthrax infection may resemble a common cold but after a few days they usually progress to severe breathing problems and finally death. Infection of persons exposed to anthrax can be prevented by early antibiotic treatment, given within hours after the exposure to the bacteria, or vaccination [5][6][7].

Botulinum Toxin: It is a muscle-paralyzing disease caused by a toxin produced by a Gram-positive bacterium called *Clostridium botulinum*. The botulinum toxin is the most toxic chemical compound known; just 0.075 micrograms can kill the average man. Symptoms of botulism will appear in 6 hours to 2 weeks and include double vision, blurred vision, nausea, muscle weakness and eventually paralysis of breathing muscles; unless assistance with breathing is provided, the infected person will stop breathing and die. If the infected person is kept breathing and alive until the antitoxin is administered to him, there is a good chance that he will eventually recover after weeks to months of supportive care (in this case only one in twenty people will die) [5][7][8].

Pneumonic Plague: It is an infectious disease caused by the Gram-negative bacterium *Yersinia pestis*, a bacterium found in rodents and their fleas. Pneumonic plague occurs when *Y. pestis* infects the lungs and usually takes one to six days to develop after infection. The first symptoms of the disease are fever, enlarged lymph nodes, headache, and cough. Without early treatment, pneumonic plague usually results in respiratory failure, shock, and rapid death. To prevent the deadly outcome of the disease, antibiotics should be given within 24 hours of the first symptoms. There is no available vaccine against plague. Pneumonic plague can be transmitted from person to person through inhalation of *Y. pestis* particles in the air [5][7][9][10].

Smallpox: It is caused by the variola virus. Smallpox was officially eradicated from the world in 1977. Currently there are only two places in the world where smallpox is held: one is the CDC center in Atlanta, Georgia, and the other is the Research Institute for Viral Preparations in Russia. Initial symptoms of the disease occur in 7-17 days and include high fever, fatigue, and head and backaches. They are followed by a rash and development of pus-filled lesions. The majority of patients recover, but death occurs in up to thirty per cent of cases. The disease can spread from person to person by infected saliva droplets. In aerosol, viruses can survive for twenty-four hours or more and they are highly infectious at even low concentrations. Administration of the vaccine within four days after infection can lessen the severity or even prevent illness [3][7][10].

3. CHEM-BIO ATTACK SCENARIOS FOR BUILDING AIR-INTAKES

We will outline two possible attack scenarios: one chemical and one biological. They both involve a person carrying the lethal material very close to the building air-intakes. The chemical attack scenario may require some short preparation on site. The biological attack scenario typically does not require any on site preparation. For modern western buildings the most deadly attack scenario is the chemical one. In the biological attack scenario the HVAC filtration system is expected to block most of the agents and minimize immediate casualties even if the building is lightly protected.

One likely chemical attack scenario to building air-intakes has already been spelled out in the testimony of Ahmed Ressam ([2]). The scenario refers to an attacker approaching the air-intakes of a building carrying a backpack and a shallow wide container. Once under the air-intakes, the attacker has to pour a liquid solution in the shallow container. The liquid solution can be kept in thermos in the attacker's backpack. Finally, the attacker drops dozens of large pills into the liquid, thus forming the chemical agent, and walks away. Wind direction and speed is very critical for the success of the attack. The way this chemical attack is planned points to Potassium Ferrocyanide that exists in pill form. When Potassium Ferrocyanide is mixed with a strong acid such as Sulfuric Acid, it creates the deadly chemical agent Hydrogen Cyanide.

Hydrogen Cyanide is lighter than air and harms or kills in medium concentrations. A lethal dose is 2,500-5,000 $mg/min/m^3$. Therefore, it renders itself as the chemical agent of choice for a "light" chemical attack. A single person can produce very quickly lethal quantities of Hydrogen Cyanide and it is highly dispersible in the atmosphere.

A biological attack scenario may involve the dispersion of a powdery substance near the air-intakes of a building. The powdery substance may have characteristics similar to weapon grade anthrax. One should expect that the natural static charge of the powdery material has been eliminated to make the agent highly dispersible. About 8,000 spores are required to infect a person. Weapon grade anthrax features 15,000 spores per mg and therefore a small amount carried by a single person can have a deadly effect, especially if the filtration system of the building is outdated.

4. ARCHITECTURE OF AN AIR-INTAKE PROTECTION SYSTEM

An effective protection system for a building's air-intakes has to be layered and geared towards prevention, early warning, and effective evacuation. A successful protection design has to take into account the architecture of the building's HVAC system. Figure 6-1 depicts a typical HVAC architecture of a modern building.

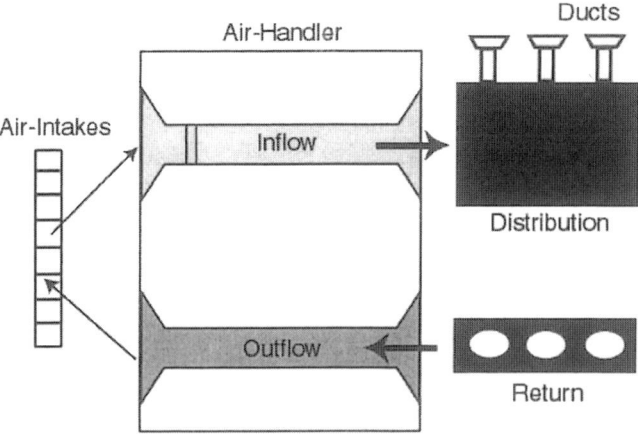

Figure 6-1. Architecture of a typical HVAC system in a modern building.

Atmospheric air flows through the air-intakes to the air-hander where it is filtered, tempered, and pumped to the distribution network of the building. From the distribution network the air flows eventually through the ducts to the building's interior living space. The critical flow path starts from the air-intakes and runs all the way to the ducts. In a chem-bio attack the warning from the protection system has to arrive in time for the pump of the air-handler to be stopped and the ducts to be blocked so that no contaminated air makes it into the building's living space.

We have run two tests with two different air-handlers from typical medium size commercial buildings to draw an idea about the reaction times required. The first test involved an air-handler that draws air from the outside at a speed of $v_n = 500$ *fpm* (*feet per minute*). The total travel path of fresh air from the air-intake until the first duct has been measured at $s_t = 110$ ft. The latency of this air-handler has been timed at $t_l = 13 \sec$. During the shut down process we assume that the reduction of air speed is linear and goes from $v_n = 500 \ fpm$ to $v_n = 0 \ fpm$ within $t_l = 13 \sec$. Therefore, the average shutdown speed is $\bar{v}_n = 250 \ fpm$. This means that during the shutdown process incoming air travels another $s_s = \frac{\bar{v}_n}{t_l} = \frac{250}{13} = 32.5 \ ft$. The worst case scenario has to assume that contaminated air goes undetected for the first $s_r = s_t - s_l = 110 - 32.5 = 67.5 \ ft$ within the air handling system at the nominal speed ($v_n = 500 \ fpm$). That means that the protection system has to issue the shutdown order to the air-handler within $t_c = \frac{s_r}{v_n} = \frac{67.5}{500} = 13.5 \sec$ from the time the agent entered the air-intake or else contaminated air will leak into the living space of the building. A similar reaction time requirement was ascertained for a different air-handler. It appears that the effective reaction time for an air-intake protection system has to be in the order of a quarter of a minute from the time the agent is released under the building's air-intakes. This is a very challenging detection requirement for a typical chem-bio sensor. As we have described in Section 2, particularly in the case of a chemical attack the terrorist has to undergo a preparation stage nearby the air-intakes before the agent is released. During this preparation time the terrorist's presence and movements are exposed to visual surveillance.

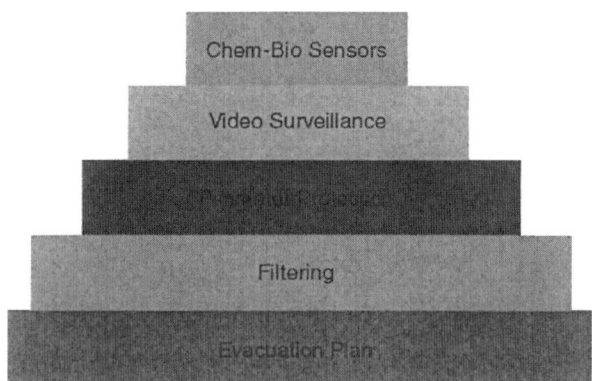

Figure 6-2. P   ' air-intakes from chem-bio attack. Layers at the bottom of the pyramid represent more basic components that have to be implemented first if the management of the building is working on a step by step approach.

6. Video-based surveillance for chem-bio protection of buildings

Assuming that the terrorist's activities will last at least several seconds in the vicinity of air-intakes an automated video surveillance system will at least double the advance warning time that any chem-bio sensor can offer. It follows from the previous analysis that video surveillance technology has to be part of an effective security solution. Our overall proposal for a layered security design for the protection of buildings' air-intakes is shown in Figure 6-2. Specifically:

Perimeter Protection: Whenever possible the area around air-intakes has to be physically protected with barbed wire or other fencing measures. This, however, is not always possible because sometimes air-intakes are over public sidewalks. This is typically the case of commercial and Government buildings in downtown areas. In suburban settings, the layout is different and air-intakes typically face an area belonging to the building campus. Fencing is far easier here because the area surrounding the air-intakes is private. The only hindrance is aesthetic. For both urban and suburban settings the most attractive case is when the air-intakes are placed at the roof of the building. Roof air-intakes are more protected by virtue of their location. They can also cater a far greater variety of preventive perimeter measures. In our view, it should be instituted that all new buildings place their air-intakes at the roof and not at their side.

Video Surveillance: While physical protection of the perimeter will make access to the vicinity of air-intakes more difficult, it will not make it impossible. An advanced video surveillance system is required to detect the presence and movements of the suspect in real-time. The purpose of video surveillance is dual: a) to provide early warning during the preparation stage of the attack and b) to record the suspect's identifiable silhouette and movements for prosecutorial purposes. The video surveillance system should feature a high degree of automation given the limited span of human attention. At the very minimum it should detect and record automatically object motion in the vicinity of the air-intakes. At the same time it should provide instant warning to the guard about the presence of the suspect. In turn, the guard should assess the situation based on the live video feed and take appropriate actions (e.g. shutdown of the air-handler). We envision a much more sophisticated video surveillance system with threat assessment capabilities. We will describe our work and vision in the next two sections.

Chem-Bio Sensors: We consider the existence of chem-bio sensors of paramount importance. Their role is also dual: a) to cross-validate the existence of a chem-bio attack signaled earlier by the video surveillance system and b) to pinpoint the type of agent released, so that appropriate decontamination measures are effected. The combination of video surveillance with an array of chem-bio sensors constitutes the pillar of the advance warning capability of our proposed protection architecture.

Filtering: Current high performance bio-filters are extremely efficient in removing bio-particulates. Buildings that are not currently equipped with such filters should consider an immediate upgrade. Although, bio filtering is of very little use in a chemical attack it can provide an effective protection in the case of a biological attack by blocking the agent in the air-handling system for a significant amount of time. Therefore, it is a preventive measure that extends the grace period within which the warning system has to respond.

Evacuation Plan: The main purpose of all the previous measures that we outlined is to provide enough time for the occupants of the building to evacuate safely. If there is no effective evacuation plan then even an on-time warning can go wasted. The evacuation plan should be hatched in advance and it should be the brainchild of a mixed group of experts encompassing facility management and security experts. The plan should be communicated to the occupants of the building and rehearsed periodically.

5. VIDEO SURVEILLANCE SYSTEM

We follow our proposed architecture in implementing the security shield for the protection of our building (Honeywell Labs Camden Building in Minneapolis, MN), which serves as our test-bed. So far, we have implemented the two bottom layers as shown in Figure 6-2: Evacuation Plan and Filtering. We have chosen not to implement the third layer from the bottom (Perimeter Protection) for aesthetic reasons. We are actively working in implementing the Video Surveillance layer. Currently our video surveillance system performs effective motion detection only. It is based on the motion detection and tracking algorithms of DETER that we have developed preciously and described in [11]. DETER has introduced a new philosophy in the design of video-based surveillance systems. Any development of the surveillance system itself is predated by a rigorous system design. The objective of the system design phase is to pinpoint the optimal number, type, and location of cameras and computational resources. The system design amounts to the solution of an optimization function (see [11]). On one hand, the optimization function ensures complete optical coverage of the surveyed area, sufficient resolution for the machine vision algorithms to perform the prescribed tasks, and enough computational power for real-time operation. On the other hand, the optimization function ensures that the technical constraints are satisfied at a minimum expense in terms of hardware units and configuration. Figure 6-3 shows the resulting optimal camera configuration for the air-intakes layout of our building (Honeywell Labs Camden building in Minneapolis, MN).

6. Video-based surveillance for chem-bio protection of buildings

We have embedded the DETER motion detection and tracking algorithm within a broader software package called DVM (Digital Video Monitor). DVM is a commercial product of Honeywell. It allows the acquisition and display of multiple video streams as well as their processing by a motion detection algorithm (DETER). Whenever sustainable motion is detected the respective incoming video stream is time-stamped and recorded on a digital storage device (hard drive). At the same time a textual and audible alarm is issue at the central console of the system to alert the operator. The DVM system has demonstrated its capability of detecting suspicious motion under difficult environmental condition in the monitoring of an oil pipeline in Central Asia for the last several months. It's performance in the monitoring of the air-intakes vicinity of our building has been equally flawless. DVM detected all the staged chem-bio attacks in our building (see Figure 6-4). At the same time, it featured almost zero false alarm rate for over a month of continuous operation (24/7). Some details about the motion detection and tracking algorithms of DETER are given in the next two subsections.

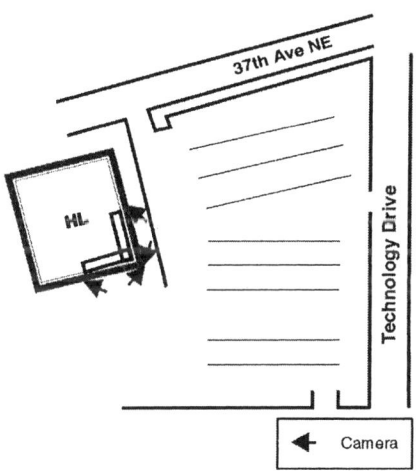

Figure 6-3. Camera configuration for the protection of the air-intakes of our building. The camera placement is the result of an optimization process. Four cameras provide full coverage in the vicinity of the air-intakes, which are placed on the side of the building and face an area belonging to our campus.

Figure 6-4. (a) Staged chemical attack. (b) Staged biological attack. Both images were captured automatically by DVM.

5.1 Motion Detection Algorithm

The DETER motion detection algorithm is based on a multi-Normal mixture model that is updated dynamically. The update mechanism is based on the incoming evidence (new camera frames). Several things could change during an update cycle:

The form of some of the distributions could change (weight π_i, mean μ_i, and variance σ_i^2). Some of the foreground states could revert to background and vice versa. One of the existing distributions could be dropped and replaced with a new distribution.

At every point in time the distribution with the strongest evidence is considered to represent the pixel's most probable background state. Figure 6-5 presents a visualization of the mixture of Normals model while Figure 6-6 depicts the update mechanism for the mixture model.

The update cycle for each pixel proceeds as follows:

First, the existing distributions are ordered in descending order based on their weight values.

6. Video-based surveillance for chem-bio protection of buildings

Second, the algorithm selects the first B distributions that account for a predefined fraction of the evidence $T: B = \arg\min\{\sum_{i=1}^{b} w_i > T\}$, where $w_i, i = 1...b$ are the respective distribution weights. These B distributions are considered as background distributions while the remaining $3 - B$ distributions are considered foreground distributions. We have experimentally established that the optimal value for threshold T is $T = 0.80$.

Third, the algorithm checks if the incoming pixel value can be ascribed to any of the existing Normal distributions. The matching criterion we use is the Jeffreys (J) divergence measure [12] and is a key differentiator of our approach from other similar approaches.

Fourth, the algorithm updates the mixture of distributions and their parameters. The nature of the update depends on the outcome of the matching operation. If a match is found, the update is performed using the method of moments [13]. This is also a key differentiator of our approach. If a match is not found, then the weakest distribution is replaced with a new distribution. The update performed in this case guarantees the inclusion of the new distribution in the foreground set, which is another novelty of our method.

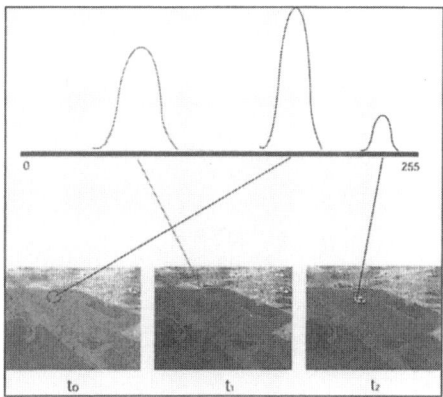

Figure 6-5. Visualization of the mixture of Normals.

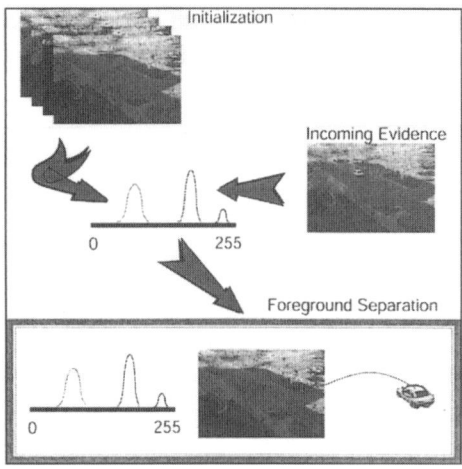

Figure 6-6. Visualization of the mixture model update mechanism.

5.2 Tracking Algorithm

In the previous section we described a statistical procedure to perform on-line segmentation of *foreground pixels* corresponding to moving objects of interest, i.e. people and vehicles. In this section, we describe how to form trajectories traced by the various moving objects. Figure 6-7 shows a snapshot of the output from the various computer vision modules of DETER. The basic requirement for forming object trajectories is the calculation of blob centroids (corresponding to moving objects). Blobs are formed after we apply a standard 8-connected component analysis algorithm to the foreground pixels. The connected component algorithm filters out blobs with area less than $A = 3 \times 9 = 27$ pixels as noise. According to our optical computation this is the minimal pixel footprint of the smallest object of interest (human) in the camera's FOV.

A *Multiple Hypotheses Tracking (MHT)* algorithm [14] is then employed that groups the blob centroids of foreground objects into distinct trajectories. MHT is considered to be the best approach to multi-target tracking applications. It is a recursive Bayesian probabilistic procedure that maximizes the probability of correctly associating input data with tracks. Its superiority against other tracking algorithms stems from the fact that it does not commit early to a trajectory. Early commitment usually leads to mistakes. MHT groups the input data into trajectories only after enough information has been collected and processed. In this context, it forms a number of candidate hypotheses regarding the association of input data with

6. Video-based surveillance for chem-bio protection of buildings

existing trajectories. Figure 6-8 depicts the architecture of our MHT algorithm.

Figure 6-7. Visualization of the computer vision operation of DETER. (a) Live video feed. (b) Segmented moving object. (c) Dynamically updated background. (d) Trajectories of the current moving objects. (e) Centroids of the moving objects. (f) Results of the blob analysis.

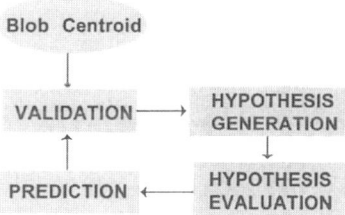

Figure 6-8. Architecture of the MHT algorithm.

6. ONGOING WORK

We are currently working on expanding the baseline air-intake system we have developed so far. We focus our efforts into two areas: a) enhancement of the video-based surveillance system and b) incorporation of additional layers of protection as described in Section 3.

The primary flaw of our current video-based surveillance system is that it does not perform any sort of substantial threat assessment. Whenever it detects motion it alerts the guard but it cannot differentiate between different types of motion. For example, the level of threat when a person passes by the vicinity of air-intakes on his way to his vehicle is trivial. But, the threat becomes significant when a person stops near the air-intakes, starts unpacking things, and finally he goes away leaving some "stuff" behind him. Therefore, we are working on designing a computer vision algorithm that will comprehend "preparation" type activities and scenes involving people leaving objects behind them.

To the question if the enhanced threat assessment system should issue a shut down order to the air-handler without waiting for the operator's assessment we respond by the development of a two-fold strategy. If the confidence from the threat assessment module is high and the environmental conditions are accommodating to a chem-bio attack, the video surveillance system will issue a shut down order before consulting with the operator or sampling the chem-bio sensors. Otherwise, the shutdown will be delayed until the operator and the chem-bio sensors weigh into the decision.

General environmental conditions in the area mean very little for this application. What matters is very localized information. This includes the wind speed and direction, the amount moisture in the air, and the temperature in the immediate vicinity of the air-intakes. The only way to continuously acquire this information in order to make real-time decisions is by incorporating into the system a computerized weather station.

ACKNOWLEDGEMENTS

We kindly acknowledge the contribution of Dr. Morellas in the tracking algorithm of DETER.

REFERENCES

[1] *Countering Biological Terrorism in the U.S.: An Understanding of Issues and Status,* Siegrist D.W. and Graham J.M. Editors, Oceana Publications, Dobbs Ferry, New York, 1999.

[2] *Witness: Terrorists Train with Poison*, Associated Press, September 26, 2001.

[3] *Threats in Bioterrorism I: CDC Category A Agents*, R.G. Darling, C.L. Catlett, K. D. Huebner, and D.G. Jarrett, Emerg. Med. Clin. N. Am. , Vol. 20, pp. 273-309, 2002.

[4] *Threats in Bioterrorism II: CDC Category B and C Agents*, G.J. Moran, Emerg. Med. Clin. N. Am., Vol. 20, pp. 311-330, 2002.

[5] *Germs. Biological Weapons and America's Secret War*, Miller, J., Engelberg, S., and Broad, W. Simon and Schuster Publications, New York, 2001.

[6] *Anthrax as a Biological Weapon, 2002. Updated Recommendations for Management*, T.V. Inglesby, et al., JAMA, Vol. 287, pp. 2236-52, 2002.

[7] *Mechanisms of Microbial Disease*, Schaechter, M., Medoff, G., and Eisenstein, B.I. Williams and Wilkins, Baltimore, 1993.

[8] *Bacterial Pathogenesis. A Molecular Approach*, Salyers, A.A. and Whitt, D.D., ASM Press, Washington, D.C., 1994.

[9] *Killer Germs. Rogue Diseases of the Twenty-First Century*, Moore, P. Carlton Books Limited, London, 2001.

[10] *Biological Terrorism: Understanding the Threat, Preparation, and Medical Response*, D.R. Franz and R. Zajtchuk, Dis. Mon., Vol. 48, pp. 493-564, 2002.

[11] *Urban Surveillance Systems: From the Laboratory to the Commercial World*, Pavlidis I., Morellas V., Tsiamyrtzis P., and Harp S., IEEE Proceedings, Vol. 89, No. 10, pp.1478-1497, October 2001.

[12] *Theory of Probability*, H. Jeffreys, University Press, Oxford, 1948.

[13] *Mixture Models Inference and Applications to Clustering*, G.J. McLachlan and K.E. Basford, Marcel Dekker, New York, 1988.

[14] *An Efficient Implementation of Reid's Multiple Hypothesis Tracking Algorithm and Its Evaluation for the Purpose of Visual Tracking*, I. J. Cox and S. L. Hingorani, IEEE Transactions on Pattern Analysis and Machine Intelligence, 18, 138--150, 1996.

II
DETECTION, TRACKING AND RECOGNITION

DETECTION, TRACKING AND RECOGNITION

Gian Luca Foresti
Department of Mathematics and Computer Science (DIMI), University of Udine, Udine, ITALY

Real-time detection, tracking, recognition and activity understanding of moving objects from multiple sensors represent fundamental issues to be solved in order to build up next generation of surveillance systems that are able to autonomously monitor wide and complex environments.

Solutions today required by the market range from standard surveillance systems that are able to detect and track objects (e.g., people, vehicles) moving in the observed scene to advanced surveillance systems based on multiple sensors/cameras (optical, infrared, thermal, radar, etc.) that are able to understand complex human behavior, to automatically detect and recognize their faces, and to discover their identity by means of specific biometric features.

Section II of the book aims at describing current research efforts in multisensor visual-based surveillance systems going in the direction of solving some of the current problems limiting a wide use of such systems. Particular attention has been focused on new techniques that are able to perform scene analysis as well as identify and track objects of interest along the field of view of multiple cameras.

This Section is divided into five contributions providing examples of research solutions currently explored by a selected set of researchers from some of the leading laboratories intensively working in the surveillance field.

In the first presented contribution, Ziliani and Reichel from Visiowave S.A., Switzerland, discuss the interaction and the integration of two important elements of most video surveillance systems: video analysis and video coding. They propose a spatio-temporal filter based on object tracking that improves coding performances and post-coding analysis. This filter seems to be an interesting alternative to complex object-based video CODECs in many video surveillance scenarios. Moreover, it increases the spatio-temporal redundancies of input signals discarding noise and unwanted objects but preserving the semantic information.

Collins, Amidi and Kanade from Carnegie Mellon University, USA, present a new solution approach to the problem of visual-based surveillance systems with active cameras. Their contribution describes an active camera system for acquiring multi-view video of moving people for applications in human identification, motion capture, entertainment and sports. A real-time tracking algorithm adjusts the pan, tilt, zoom and focus parameters of multiple active cameras to keep the moving person centered in each view. The output of the system is a set of synchronized, time-stamped video streams, showing the person simultaneously from several viewpoints.

The third contribution is presented by Amer from Concordia University, Canada. A computational layered framework to extract multiple high-level features of a video shot is presented. The objective with this framework is to extract rich high-level video descriptions of real world scenes. In the proposed system, high-level descriptions are related to moving objects, and moving objects are represented by their quantitative and qualitative low-level features. Semantic features are then represented by generic high-level object features such as events. To achieve higher applicability, descriptions are extracted independently of the video context.

Four interacting video processing layers have been developed: (a) enhancement to estimate and reduce noise, (b) stabilization to compensate for global changes, (c) analysis to extract meaningful objects, and (d) interpretation to extract context-independent semantic features. Extensive tests on indoor and outdoor environments in the presence of multi-object occlusion, noise, and artifacts have been presented.

Porikli from Mitsubishi Electric Research Laboratories presents a new method for automatic object tracking and video summarization for multi-camera systems with a large number of non-overlapping field-of-view cameras is explained. In the presented system, video sequences are stored for each object as opposed to storing a sequence for each camera. Object-based representation enables annotation of video segments, and extraction of content semantics for further analysis and summarization. Objects are tracked at each camera by background subtraction and mean-shift analysis. Then, the correspondence of objects between different cameras is established by using a Bayesian Belief Network. This framework empowers the user to get a concise response to queries such as "which locations did an object visited on Monday and what did he do there?".

Detection, tracking and recognition 117

In the last contribution Foresti from the University of Udine, Italy, and Giacinto and Roli from the University of Cagliari, Italy, present an advanced visual-based surveillance system that is able to detect and track in real-time multiple moving objects and at the same time learn and recognize their behavior in complex outdoor scenes, i.e., parking areas, where a human or a vehicle can assume several and different behaviors. The number of possible events (normal and/or dangerous) that occur in the observed scene is very high. Consequently, they propose the use of recognition techniques based on a "learning by example" paradigm in order to reduce the complexity of the learning process from real situations. A statistical model of the distribution of "normal" and "dangerous" trajectories based on a relatively small set of examples (a set of "prototypal" trajectories manually selected), has been derived. Then, the event recognition problem has been formulated in terms of a "pattern recognition" problem, where the trajectories, represented by a Beziér curve, play the role of "patterns" that must be assigned to one of the two data classes, e.g. "normal" or "dangerous" trajectories. Finally, the whole set of sample trajectories is stored into an event database and used to train a multi-layer perceptron (MLP) neural network.

Chapter 7

SECOND GENERATION PREFILTERING FOR VIDEO COMPRESSION AND ANALYSIS IN MULTISENSOR SURVEILLANCE SYSTEMS

Francesco Ziliani and Julien Reichel
Visiowave S.A., Switzerland

Key words: Second-generation video prefiltering, video compression, video analysis, video surveillance

1. INTRODUCTION

Large digital surveillance networks that are deployed on buildings, highways, cities, train and metro stations, integrate an increasing number of cameras and sensors. Human operators cannot monitor all the available visual data without reliable and automatic detection systems. This is why object detection and tracking techniques, that are the first important steps towards higher level of scene understanding and analysis, become critical in video surveillance scenarios.

In the same context, video coding plays a fundamental role. Video coding solutions must provide the best rate-distortion ratio compatible with strong constraints on the coding-decoding delays, on the cost of the solution and on its added functionalities. High compression performances are required to save the available network bandwidth and to increase the storage capabilities. Delays constraints are introduced to enable a fast interaction between the observer and the scene for example in the case of manual remote control of mobile cameras. The complexity of the video coding solutions is limited by the hardware and its cost. Functionalities such as spatio-temporal scalability, fast forward and backward and error resilience are additional requirements in most surveillance scenarios. All these constraints influence the choice of the most suitable CODEC and often proprietary solutions are engineered to optimize all the compromises.

Despite the high number of possible solutions, any video coding algorithm achieves the compression of an input signal by exploiting its spatial and temporal redundancies. The higher is the spatial and temporal homogeneity of the input signal, the higher will be the compression factor. In the domain of security, the input signal is often acquired by fixed cameras. In these circumstances the temporal redundancy of the data is theoretically extremely high and should be easy to compress. However, in real scenarios, the moving objects in the scene, the illumination drifts and the camera noise reduce the temporal homogeneity of the signal. The visual effects introduced by most of these phenomena are extremely expensive to code due to their low spatio-temporal redundancy. Moreover, in most security applications, effects such as illumination drifts, camera noise, the movement and the texture of objects that are not of interest for surveillance purposes (e.g. the movement of tree branches) are not significant in improving the understanding or the analysis of the observed scene for a human observer or an automatic event detector.

In the above context, the concept of second-generation video coding introduced by Kunt in [1] is extremely appealing. In second-generation video coding, the video sequence is interpreted as a set of perceptual features. Some features might be more important then others in improving the perceptual quality of the scene and thus they need to be preserved during a lossy coding process. In a general context an example of perceptually significant features are the edges of the scene. In other applications such as video surveillance a feature could be an entire region of interest (ROI).

In the video coding research community, the above observations suggested the use of object-based video coding such as the H.263+ Video CODEC [2]. The idea behind object-based video coding is to preserve the quality of the regions of interest (ROI), by allocating to them a higher bandwidth. This is done by modulating the quantization coefficients in the image: the blocks corresponding to the ROI will be quantized less than the blocks that do not belong to ROI. By distributing more bandwidth to the important parts of the scene the visual quality is improved.

This solution is certainly attractive, but has a number of limitations in the context of video surveillance. First, it requires object-based enabled CODECs and not all existing solutions guarantee this feature in particular in the domain of video surveillance where the compatibility with already installed systems is often required. Moreover, existing object-based video CODECs are complex schemes that do not always fit to the strong cost, delays and functionality requirements of real surveillance applications.

In existing object-based video CODEC standards, the objects of interests have to be described by masks, preferably blocky [3] to avoid that the shape-coding overhead reduces the available bandwidth. Blocky masks reduce the contour accuracy of detected objects. In addition to these drawbacks, the visual quality of the scene that has not been detected as ROI may be extremely degraded to compensate for the additional overhead introduced by the need of sending the mask information to the decoder. The artifacts introduced in these areas reduce the reliability of further object detection and tracking systems.

The approach we propose in this contribution is different, although based on similar assumptions. Instead of concentrating on modifying the coding algorithm that must satisfy already several constraints, we apply the concept of second-generation video coding to the prefilter stage. We will show that this choice has a number of advantages in particular in the context of video surveillance.

2. SECOND-GENERATION VIDEO PREFILTERING

Prefiltering is well known in video coding to reduce the noise of video sequences and to correct several kind of defects such as scratches or image vibrations. Here we introduce a second-generation filtering approach, where the filtering stage simplifies the input sequence according to a specific definition of what are the relevant image features. In order to reduce the encoding cost of unimportant image features, it is necessary to increase their spatial and or their temporal redundancy. This can be obtained with linear or non-linear low pass filters. The concept of second-generation filtering may be applied to any video coding solution as far as it is possible to define the relevance of image features.

In the context of video surveillance, the objects moving in the static scene are in general relevant image features. If it is possible to detect and track the moving areas in the scene and decide, according to their spatio-temporal properties their priority for a given application, it is possible to distinguish between areas of interest and areas of non-interest. The areas of non-interests are "smoothed" while the ROI are let unchanged. The result is a scene where the ROI remains at the original quality, while the rest of the scene is easier to encode.

In this context we propose an Object Tracking-based prefilter in order to modify the input signal and increase its spatial and temporal redundancy. The filter we propose aims at recovering the spatial and temporal redundancy of the input signal by minimizing the coding cost of illumination drifts, camera noise and unimportant moving objects. The mechanisms of the proposed filter are simple. First, a simple, yet effective object detection and tracking step is performed on the original signal. Then, the video signal is temporally filtered taking into account the results of object tracking.

The advantages of the proposed system can be summarized as follows. The object detection and tracking is done on the original input sequence reducing the chances of misdetection introduced by coding artifacts. The obtained filtered signal is extremely simple to code: only objects of interests have a significant coding cost. The coded signal has a high visual quality on the whole scene that is preserved during coding because of the high temporal uniformity. Thus the coded signal can be efficiently used as input to object tracking systems. The proposed prefilter can be applied on whatever video CODEC.

In the following we describe the mechanisms of the proposed scheme. First we introduce the adopted object tracking algorithm and its advantages in the context of our solution. Then we define the temporal filter based on the object tracking results. Finally we demonstrate the effects of the filter on standard and non-standard video CODECs.

3. OBJECT TRACKING AS PREFILTER

The proposed filter aims at increasing the temporal redundancies of the input signal. This is done in two steps as detailed in Figure 7-1. First an object-tracking algorithm is applied on the original signal to define the ROI present in the scene. Then, all the areas that do not correspond to the ROI are temporally filtered.

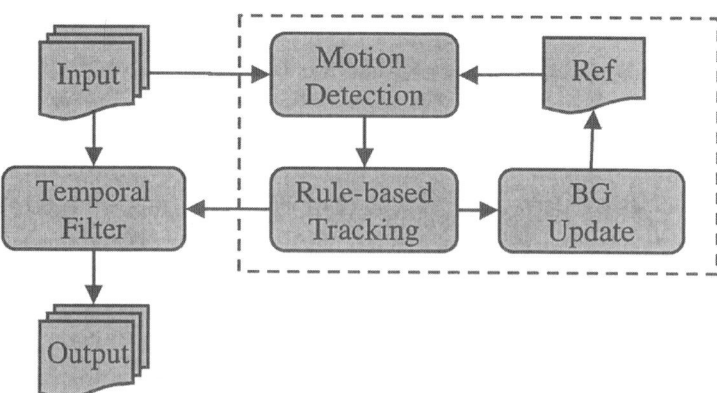

Figure 7-1. Scheme of the proposed Object tracking prefilter. The input sequence is first used by the Object Tracking module to identify the objects of interest and their properties. Then the Temporal Filter simplifies the input according to the results of the Object Tracking.

The filter is currently designed for video surveillance applications where a fixed camera controls an indoor or outdoor scene. In this context the ROI can be simply defined as the *moving objects* in the scene such as cars, persons, animals etc...

The detection of the ROI is performed with a three steps algorithm that satisfies low computational constraints. First, a Motion Detection procedure has been developed to detect all the moving areas in the image by comparing a reference frame with the current one. Then, these areas are tracked by solving their temporal correspondence from frame to frame. The temporal correspondence is used to understand the temporal properties of each moving areas and to label them as: spurious region, new appeared region, still region, moving region. Finally, the results of the temporal correspondence are used to update the reference background and to filter the input signal.

3.1 Motion Detection

The first step of the proposed filter is motion detection. Motion detection is used to find in the image all the ROI candidates areas[1]. There exist several techniques to detect motion in images [4]. We have selected an approach that satisfies our low complexity and high accuracy criteria: the model-based Statistical Change Detection [5]. This approach is based on the definition of a simple statistical model of the image background. The validity of the model is tested at each image position. Following the suggestions proposed in [2],[6] we have defined an improved version of this algorithm. The input is represented by three luminance images: the reference $R(t)$, the current image $I(t)$, and the next image $I(t+1)$. For each pixel position k in the image, a 5×5 neighborhood η_k centered in k is defined. A 5×5×2 spatio-temporal cube Ω_k is defined by computing at each position in η_k the pixel-wise difference d between $R(t)$ and $I(t)$, and between $R(t)$ and $I(t+1)$. Ω_k summarizes in a given image position, the spatio-temporal properties of the scene. If the pixel k lies on the background of the scene, all the terms in Ω_k should be zeros. In reality, since noise affects each frame of the input signal, all the elements of Ω_k are well approximated by statistical independent random variables belonging to a zero-mean gaussian distribution $N(0, \sigma)$. The higher is the noise affecting the video signal the higher is the σ. This parameter can be measured offline and characterizes the properties of the

---

[1] We assume that in surveillance applications all the regions of interest (ROI) are or have been moving objects.

acquisition system [6]. Under these hypothesis the following random variable:

$$\Delta_k = \frac{\sum_{\Omega_k} |\Omega_k|}{\sigma} \qquad (7.1)$$

belongs to a χ^2 distribution with *100* degrees of freedoms (twice the number of elements in Ω_k) [7]. The probability that the observed Δ_k will exceed the value χ^2 by chance even for a correct model is given by

$$Q(\chi^2|\eta) = 1 - P(\chi^2|\eta) = 1 - P\left(\frac{\eta}{2}, \frac{\chi^2}{2}\right) \qquad (7.2)$$

where η is the number of degrees of freedom of the selected χ^2-distribution and $P(\cdot,\cdot)$ is the incomplete gamma function [8]. Given Equation (7.1) all the pixels k belonging to the background are defined by the following test:

$$Q(\Delta_k|100) < t_\alpha \approx 0.001 \qquad (7.3)$$

The remaining pixels belong to the foreground and are defined to be the moving objects in the scene. The definition of the spatio-temporal cube Ω_k improves the statistical and the temporal coherence of the original approach proposed in [6]. The obtained detection mask does not require further refinements with morphological operators.

3.2   Temporal Correspondence

As soon as motion detection detects moving areas in the scene, a connect component analysis is applied to the corresponding binary mask so as to define a set of spatial disconnected blobs. Each blob is a new *father* region and it is associated to a single label. If these regions correspond to moving objects in the scene, the motion detection will detect moving areas in the following frames as well. The object tracking algorithm aims at defining the correspondence between the father regions and the newly detected moving areas (*children*). This is achieved in three steps.

The first step exploits only the information on the relative position in the image between fathers and children. The simplest case is the **perfect correspondence**: a single father covers a single child and vice-versa. Then we define the following three alternatives cases: **no correspondence**,

7. Second generation prefiltering for video compression and analysis in multisensor surveillance systems

multiple correspondence and **exclusive correspondence**. The first happens when a father does not cover any children. The second is when more than one father cover the same child. The third is when a father covers more than one child and all these children are covered only by that father.

Perfect correspondence solves in an extremely fast way all simple but important situations where a single moving object has smooth trajectories and is not disturbed by adjacent objects. The remaining cases require further treatment, performed in step two. Here we treat fusion and occlusion by taking into account region similarities based on mean color and movement.

In the last step the remaining father and children regions that have not found valid correspondence in the previous steps, are respectively defined disappeared and new regions.

3.3 Updating the Reference Background

Once the temporal correspondence is solved, it is possible to access to the temporal properties of all the detected blobs. Based on these properties we classify them in one of the following object classes: Background, New, Young and Old object classes. The Background class collects all the pixels that have not been detected as changed by the motion detector. The New class collects all the pixels belonging to blobs that have not found a correspondence with the blobs detected in the previous frame. The Young and Old classes collect all the pixels belonging to the blobs that have found correspondences respectively below and above a consecutive number A (age) of frames. A typical value for A is 10 frames. The reference background is updated differently according to how the pixel k is classified in one of the above classes. The equation we propose is the following:

$$R(t,k) = [1-\alpha(k)] \times R(t-1,k) + \alpha(k) \times I(t,k). \tag{7.4}$$

where $R(t,k)$ and $I(t,k)$ are the intensity values of the same pixel position k at time t, respectively in the reference and the input image and $\alpha(k)$ is a real value between 0 and 1. The weighting factor $\alpha(k)$ controls how fast the reference image is updated with the input images. For example, if $\alpha(k)$ is 1/25, it will be necessary to wait 25 frames (equivalent to 1 second in PAL format) to reflect in the reference image the precise change happened in the scene. When updating the reference background we would like to satisfy the following:
1. the progressive illumination changes of the scene should be reflected in $R(t,k)$ so as avoid false detection of moving areas when compared to $I(t,k)$.

2. the moving objects in the scene should not affect $R(t,k)$ so as limiting the effects of covered and uncovered background.
3. new objects appearing in the scene should update $R(t,k)$ so as not to leave ghost blobs.
4. areas detected as changed but for which it is not possible to find stable temporal correspondences are probably moving areas of not interest and $R(t,k)$ should be updated so as inhibit their detection.

In order to take into account the above observations, we propose the following technique to define $\alpha(k)$. First, we check if the pixel k has been classified in the Slow class in the last P frames: in this case a persistency mode is activated and $\alpha(k)$ is fixed to 1/P. A typical value for P is 10. Otherwise α_k is defined according to Table 7-1.

Table 7-1. Proposed values for the weighting factor $\alpha(k)$.

Pixel k classified as:	Value for $\alpha(k)$
Background	1/25
New	0
Young	1/50
Old	1/75

The procedure described above can be intuitively understood as follows: all the pixels that have been classified as background are updated slowly so as to take into account for all the global illumination changes. The detection of a new object freeze the reference background. This has two advantages: avoid the detection of spurious objects that vibrate or move in a periodic way over a small area, and increase the detection capabilities for objects that have a real trajectory (the increased temporal distance increases the changes and eases their detection). If the object is not new, but its temporal correspondence has not been solved robustly (at least for A consecutive frames), then the background should not be polluted too much by the values of the moving object (low integration speed: 2 seconds). Moreover, a persistency mode is activated in this case to recover the situation where a still object originally present in the background start moving: in this case, its original position will be rapidly updated ($\alpha(k) = 1/P$, with P = 10). Finally if the object is detected robustly enough for at least A consecutive frames, it is possible to further reduce the pollution of the background with the intensity values of the object (very low integration speed: 3 seconds). This improves the detection of the whole object, by reducing the problem of covered and uncovered background.

Note that the values proposed in Table 7-1, may be easily adapted to the constraints of a specific application by changing the numerical values associated to the different classes. In specific scenarios, it could be imagined to add additional classes with specific integration speeds and mechanisms.

3.4 Temporal Filtering the input image

A similar approach used to update the reference background is used to temporal filtering the input image. The output at time *t* and at the pixel position *k*, is a function of the input at time *t* and the output at time *t-1*.

$$O(t,k) = [1-\alpha(k)] \times O(t-1,k) + \alpha(k) \times I(t,k) \tag{7.5}$$

O(t,k) is the intensity value of a pixel position *k* at time *t*, in the output image and α(*k*) is a real value between 0 and 1. The temporal filtering is responsible for the simplification of the input signal so as to increase its temporal redundancy. In particular it is desirable to reduce the camera noise, the image vibrations and to smooth the unimportant moving objects, such as moving trees, while it is important to preserve the temporal resolution of the objects of interest. In order to achieve these objectives we propose to select an appropriate α(*k*) according to the classification of the pixel position *k* in one of the object classes defined in the previous section. The values evaluated in the experimental results are displayed in Table 7-2.

Table 7-2. Proposed values for the weighting factor α(k).

Pixel *k* classified as:	Value for α(*k*)
Background	1/10
New	1/2
Young	2/3
Old	1

As commented in the previous section, the values reported in *Table 7-2*, may be easily adapted to the properties of the input sequence. In particular, the value proposed for the background could be computed according to an estimation of the thermal noise introduced by the camera system. The objects classified in the New and Young classes are temporally filtered to take into account the fact that they may contain spurious objects. A visual inspection of the results of the proposed filter clearly show a reduction of the camera noise and image vibrations introduced by these values. In the following section, a more detailed report of the results in terms of coding performances is reported.

4. RESULTS

The proposed filter has been tested on a number of public available surveillance sequence (COST and PETS[2]). These sequences show two typical surveillance scenarios: surveillance of highways and surveillance of outdoor parking. In all situations, the camera is fixed and affected by camera noise. The filter has been evaluated on 3 criteria. The ability to detect and track moving objects in the scene.

The improvement of the compression-ratio compared to other solutions and the improvement of the detection and tracking results on the compressed stream. The results have demonstrated that although the proposed solution is extremely simple, all the moving targets in the field of view have been successfully detected.

The object tracking has been correctly solved in regular situations, when no occlusions or crossing were happening. It fails occasionally when complex situations such as multiple occlusions and disocclusions arise. This result was expected since no special handling of occlusions have been defined. The misclassifications caused by the above limits of the suggested implementation have a marginal effect both on the perceptual quality of the filtered image and on the measured coding gain.

Figure-7-2: Example of tracked object for the sequence n25w10 from COST database. The trajectory of the object 11, tracked for 290, frames is displayed.

---

[2] COST database is available at http://www.tele.ucl.ac.be/EXCHANGE/, PETS2001 database is available at ftp://pets2001.cs.rdg.ac.uk.

7. Second generation prefiltering for video compression and analysis in multisensor surveillance systems

Figure 7-3: Example of tracked objects for a sequence from the PETS database. The trajectories of the two objects 67 and 55 respectively tracked for 409 and 87 frames are displayed.

In Figure 7-2 and Figure 7-3 we represent the trajectories of few targets, detected and tracked in the scene.

In order to evaluate the influence of the proposed filter on video coding, we have selected two different CODECs. The first CODEC is based on a 3D wavelet transform of the input signal without motion prediction. This solution is currently used in professional security surveillance systems and offers a good compromise between computational complexity, functionalities and compression ratios. In order to guarantee a fast random access to the sequence, this CODEC introduces several intra frames. This limits its efficiency in exploiting the temporal redundancy of the input sequence. The second CODEC is an MPEG-4 based solution with motion prediction and GOP length of 300 frames. This CODEC offers much better rate/distortion performances with higher complexity and less functionalities.

Both CODECs have been configured to encode at fixed quality the input signal. Note that fixing the quality is equivalent to fix the quantization tables. Then, the original input sequences and their filtered version have been compressed with the two CODECs.

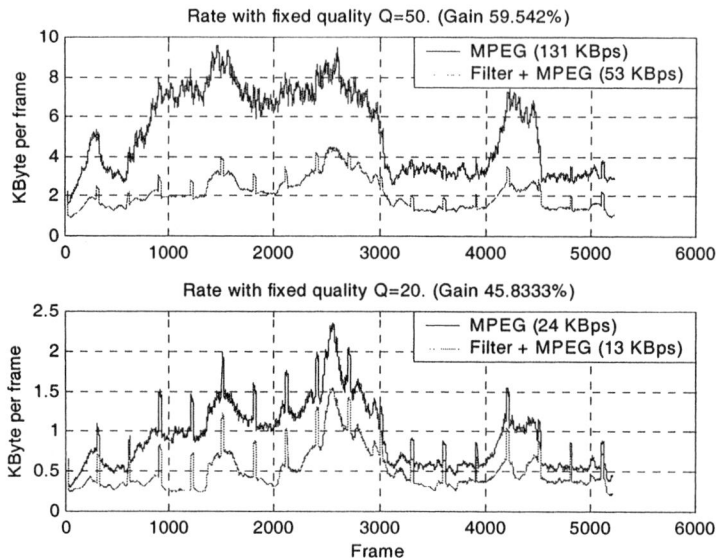

Figure 7-4: Comparison between the rate obtained on the original sequence and the filtered one by using an MPEG CODEC with fixed qualities Q=50 (top) and Q=20 (bottom).

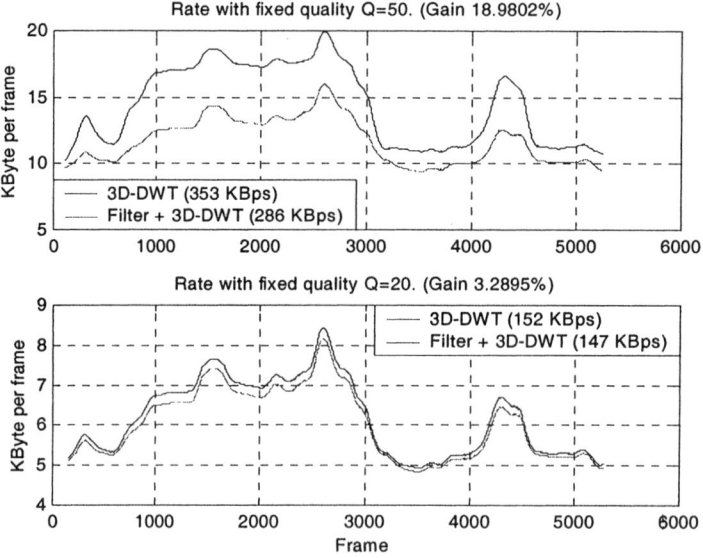

Figure 7-5: Comparison between the rate obtained on the original sequence and the filtered one by using a 3D-DWT-based CODEC with fixed qualities Q=50 (top) and Q=20 (bottom).

A representative example of the obtained results is reported in Figure 7.4 and Figure 7.5 for a single sequence of the PETS database in CIF format at 25 frames per second.

The obtained results confirm that the proposed filter improves the performances of both CODECs and at different coding rates. In the results reported in Figure 7.4 we measured an improvement of 59.5% and 45.8% for the MPEG CODEC with quality Q=50 and Q=20. These quality factors provided a distortion of 45 dB and 38 dB respectively. In the results reported in Figure 7.5, we measured a less important improvement of 18.9% and 3.2% for a 3D-DWT CODEC with quality Q=50 and Q=20. These quality factors provided a distortion of 40 dB and 37 dB respectively. The less important improvement is due to the limited use of the temporal redundancy performed by this CODEC.

In addition to the gain in the rate/distortion performance of the CODECs, the visual inspection of the decoded results underlined the contribution of the filter in improving the overall perceptual quality of the scene and in reducing the coding artifacts such as ringing and mosquito noise. The major advantages have been obtained when the sequence contained high temporal noise and disturbance moving areas such as trees.

An interesting result of the proposed temporal filter is that the background areas are extremely stable and temporally similar. Consequently, from the decoded stream, the background areas remains temporally uniform. This can be exploited to detect the background areas at the decoder with a simple thresholding of the compressed image difference.

5. CONCLUSIONS

We have introduced the concept of second-generation prefiltering. This is a flexible approach that simplifies the input sequence by taking into account the relevant image features. This simplification has the aim of improving the rate-distortion performances of any CODEC scheme without degrading the image features of interest. Relevant image features may be perceptual features such as the edges or more application oriented image features such as region of interest (ROI).

We have applied the concept of second-generation prefiltering in the context of video surveillance. Here, relevant features are the regions of interest. In this framework we have proposed a temporal filter based on the image analysis results obtained by a simplified object tracking system. The filter aims at removing from video surveillance sequences all the unimportant visual information such as spatial and temporal noise, slow illumination drifts and incoherent moving objects.

The visual quality of the filtered sequence is preserved, while its spatio-temporal redundancies are maximized. This dramatically improves the performances of successive video CODECs. Tests have been performed on public available video surveillance sequences and on commonly used video CODEC systems with consistent improvements in video compression ratios. Moreover, from the decoded signal it is extremely simple to extract the areas that were detected as moving objects at the encoder side.

The results suggest that the proposed prefilter is a valid alternative to complex object-based video CODECs in several video surveillance scenarios.

REFERENCES

[1] M. Kunt, A. Ikonomopoulos, and M. Kocher, "Second generation image coding techniques," *Proc. IEEE*, vol. 73, no. 4, pp. 549–574, April 1985.

[2] G. L. Foresti, P. Mähönen, and C. S. Regazzoni, *Multimedia video-based surveillance systems*, Kluwer Academic, 2000.

[3] F. Moschetti, G. Covitto, F. Ziliani, and A. Mecocci, "Automatic object extraction and bitrate allocation for second generation video coding," in *Proceedings of ICME 02*, Lausanne, Switzerland, Aug. 2002.

[4] A. Mitiche and P. Bouthemy, "Computation and analysis of image motion: A synopsis of current problems and methods," International *Journal of Computer Vision*, vol. 19, no. 1, pp. 29–55, 1996.

[5] T. Aach, A. Kaup, and R. Mester, "Statistical model based change detection in moving video," *Signal Processing*, vol. 31, no. 2, pp. 165–180, Mar. 1993.

[6] F. Ziliani and A. Cavallaro, "Image analysis for video surveillance based on spatial regularization of a statistical model-based change detection," in *Proc. of the 10th International Conference on Image Analysis and Processing*, Venezia, 1999, pp. 1108–1111.

[7] T. Aach, A. Kaup, and R. Mester, "Statistical model based change detection in moving video," *Signal Processing*, vol. 31, no. 2, pp. 165–180, Mar. 1993.

[8] W. H. Press, S. A. Teukolsky, W. T. Vetterling, and B. P. Flaunery, *Numerical Recipes in C*, Cambridge, 1993.

Chapter 8

ACQUIRING MULTI-VIEW VIDEO WITH AN ACTIVE CAMERA SYSTEM

Robert T. Collins, Omead Amidi, and Takeo Kanade
Carnegie Mellon University

Key words: Multi-view stereo, active tracking, video surveillance, sports video

1. INTRODUCTION

For applications in human identification, activity recognition, 3D reconstruction, entertainment and sports, it is often desirable to capture a set of synchronized video sequences of a person from multiple camera viewpoints (see Figure 8-1). One way to achieve this is to set up a ring of cameras all statically aimed at a single point in space, and to have an actor perform at this fixation point while the video footage is shot. This is the method used to create spectacular special effects in the movie The Matrix, where playing back frames from a single time step, across all cameras, yields the appearance of freezing the action in time while a virtual camera flies around the scene. However, in surveillance or sports applications it is not possible to predict beforehand the precise location where an interesting activity will occur, and therefore it is necessary to dynamically adjust the fixation point of multiple camera views. We have developed a system that tracks a person in real-time and adjusts the pan, tilt, zoom and focus of each camera to acquire synchronized multi-view video of a person moving through the scene.

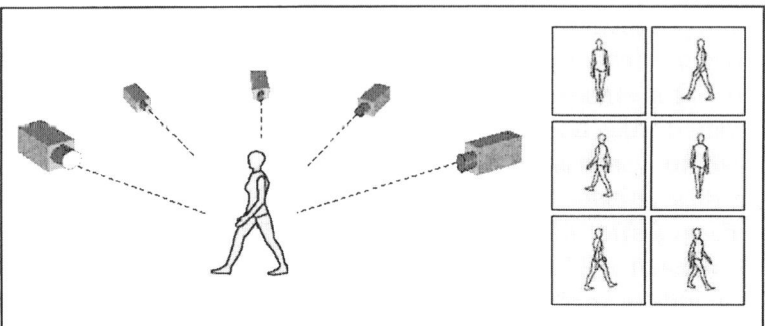

Figure 8-1. Multiple active cameras are used to acquire synchronized views of a moving person from multiple viewpoints.

2. SYSTEM DESIGN

The system design emphasizes modularity of each individual camera by minimizing the amount of information each camera has about the other cameras. Such modularity is useful in practical situations since all cameras may not necessarily be installed at the same time -- in fact, new cameras may be added, and some may malfunction and be removed, even while the rest of the system is running. To achieve this level of modularity, each camera is calibrated independently with respect to a 3D scene coordinate system, and intercamera relationships are only implicitly represented through their joint individual relationships with the 3D scene. The cameras communicate by passing geometric "messages" through the shared 3D scene geometry. For example, there is no explicit representation of the epipolar geometry between any pair of cameras. A viewing direction from camera A is transformed into a 3D oriented ray in scene coordinates, which when projected into camera B essentially defines the appropriate epipolar ray. In this way, each camera can be treated independently, leading to a system in which truly distributed multi-camera processing can occur.

Figure 8-2 shows a block diagram of one active camera module. A computer CPU is connected to a camera mounted on a pan/tilt device. This CPU is responsible for processing video from the camera, and based on the results it adjusts pan, tilt, zoom and focus parameters to maintain tracking, e.g. to keep the tracked person centered in the image. The camera is synchronized to a common system-wide genlock signal, so that the shutter for each camera fires at precisely the same time, resulting in video frames taken at the same time instant. Video from the camera is time-stamped using a VITC time code generator that inserts a system-wide time stamp directly into the vertical blanking interval of the video signal. This video branches both to the tracking CPU, and to a hard drive for recording. The tracking CPU communicates with other camera modules through a local area network.

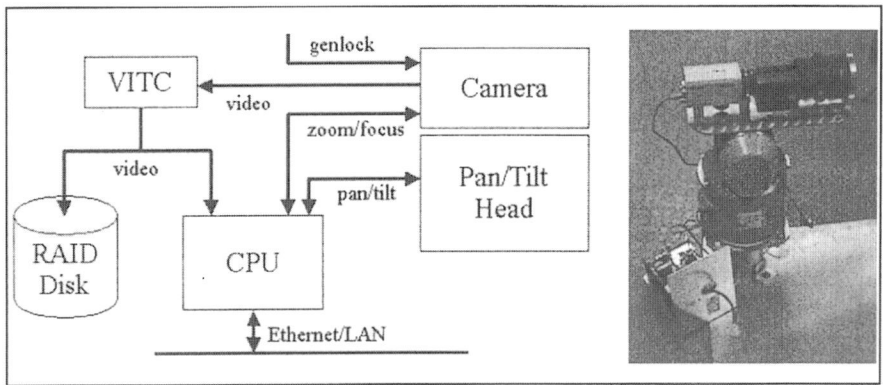

Figure 8-2. (Left) Block diagram of a single camera module for actively acquiring time-stamped video of a moving person. (Right) Prototype hardware implementation of this module.

Figure 8-2 also shows a hardware prototype of this camera subsystem. The camera body is a Sony DXC-950, 3 CCD color camera that produces interlaced NTSC video output. To that is mounted a Canon YH18x6.5 motorized zoom lens. At high zoom, this lens has approximately a four degree field of view. The pan/tilt head consists of the first two joints of a Mitsubishi Heavy Industries industrial robot arm. This head was chosen for its accuracy, repeatability, and ability to carry a moderately heavy payload. A small industrial computer running the VxWorks operating system provides real-time image processing and camera control. Not shown in the picture is a Linux PC containing the RAID disk to which full-frame color video is streamed at 30 frames per second.

3. ACTIVE TRACKING ALGORITHMS

Each camera module actively adjusts its pan, tilt, zoom and focus to track a target person in real-time, as described below. Since single-camera tracking is sensitive to occlusions and clutter, the set of camera modules communicate to fuse their individual location estimates into a 3D estimate of the person's location in the scene. Cameras that lose track of the person can thus recover, as long as some subset of cameras has continued to correctly track the person.

3.1 Single-camera tracking

Much previous work in people detection and tracking for surveillance uses adaptive background subtraction [1] [2] [3] [4]. However, in the current system it is necessary to track a moving person while the camera is panning, tilting and zooming. Although theoretically video from a rotating and zooming camera can be registered and subtracted from a panoramic mosaic [5], adaptation to lighting changes is difficult since the whole panoramic scene is not being viewed continuously, and there are issues in representing panoramas at multiple scales when variable camera zoom is present.

To track objects from a continuously moving camera, we use the mean-shift algorithm [6] [7]. Each pixel in a window of interest within the incoming video frame is assigned a likelihood of belonging to the person being tracked, using an appearance model learned when the person is first sighted. This likelihood map represents an implicit probability distribution on the location of the person in the 2D video frame. The mean-shift algorithm is a non-parametric method for rapidly finding the nearest local mode of this distribution (Figure 8-3). The 2D location found is used to control the camera pan and tilt parameters to keep the person in the center of the image.

We currently use a simple appearance model based on a histogram in normalized color space of the pixel colors falling within a rectangle centered on the person. Future work will improve the appearance model to incorporate multiple cues including texture, shape and predicted motion.

Figure 8-3. Two-dimensional tracking is achieved by applying the mean-shift algorithm to an image where pixel values represent likelihood of belonging to the tracked person. The mean-shift algorithm is a non-parametric method for climbing to the nearest local mode of this likelihood map.

3.2 Multi-camera location updates

Camera modules communicate to achieve a consensus on the person's 3D location. This allows cameras that have a good view of the person to aid other cameras with poor or occluded views. Once per second, each camera i broadcasts a time-stamped UDP packet containing its 3D focal point location c_i its principal viewing ray orentation u_i computed from current pan and tilt angles, and a weight w_i that specifies how confident the camera is in its current tracking results. This weight w_i could be the inverse angular variance in pointing direction to the object from that camera, or some other confidence measure. If the camera has lost track of the object, it can set its weight to zero.

The message from each camera i constrains the person to lie along a 3D ray $c_i + k\, u_i$, with k being a positive distance along the ray. When two or more cameras view the same person, an estimate of the person's 3D location can be computed via triangulation of these viewing rays (Figure 8-4). Due to inaccuracies, these rays will not intersect exactly at a single point, but we can compute a *pseudo-intersection* point P that minimizes the sum of squared distance to each pointing ray. Point P is found as the solution to the linear system of equations:

$$\left[\sum_i w_i \left(I - u_i u_i^T\right)\right] P = \sum_i w_i \left(I - u_i u_i^T\right) c_i \qquad (8.1)$$

This 3D estimated location P is used to adjust the pan/tilt angles of each camera, to an extent that depends on the camera's tracking confidence w_i. Cameras from which the person is occluded can therefore continue to track the virtual position of the person from their viewpoint.

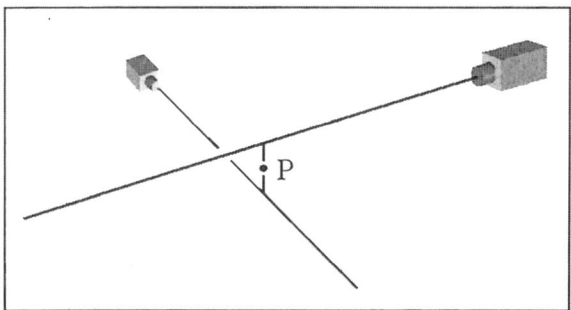

Figure 8-4. Multiple camera viewing rays are fused into a single object location estimate by finding the pseudo-intersection point P that minimizes sum of squared distance to each pointing ray.

The distance from estimated 3D location P to each camera's location c_i can be used to control camera zoom and focus to keep the person the same size and in sharp focus in all the images, even though the cameras are different distances away from the person. Let r be the desired radius of a virtual sphere subtending the entire vertical field of view of each image. Let $d_i = sqrt((P - c_i)^T(P - c_i))$ be the distance between camera location c_i and point P. Then the desired vertical field of view angle α can be computed as $\alpha = 2\ arctan(r / d_i)$. The zoom parameter that achieves this desired field of view is then computed from a lookup table. When using lenses that do not have autofocus, the focus lens parameter for distance d_i is also computed for each camera, using a pre-calibrated lookup table.

4. CAMERA CALIBRATION

Before operation of the system, each camera is calibrated so that its relationship to the scene is explicitly known. This requires determining the pose (location and orientation) of the camera with respect to a scene coordinate system, determining the relationship of the zoom control parameter to angular field of view, and determining the relationship of the focus control parameter to the distance of objects in the scene.

Camera pose is determined by measuring pan/tilt angles towards a set of distinguished points or "landmarks" with known 3D coordinates. The 3D landmark points are determined prior to calibration by surveying with a GPS unit or theodolite. Sighting each landmark involves rotating the pan/tilt device from a user interface, until the landmark point is centered within the field of view of the camera. The pan/tilt parameters at this position are then stored with the X, Y, and Z coordinates of the landmark, to form one pose calibration measurement. Camera orientation R and location c are determined by an optimization procedure that minimizes the angle between pan-tilt viewing rays rotated by R and direction vectors from the camera origin c to the 3D landmark points. The basic pose solution method is presented in [8].

For high-precision pointing, it is also necessary to measure the pitch and yaw of the sensor as mounted on the pan/tilt device, and the offset of the sensor focal point from center of rotation of the pan/tilt device. In practice, we measure the offset by hand using a ruler, and solve for the pitch and yaw by embedding the basic pose solution method inside an outer optimization loop that searches for the pitch and yaw angles that minimize residual pose error.

Computer control of motorized zoom lenses involves sending the desired zoom and focus as a command to the camera/lens system. The effect of the value of these parameters on physical lens settings must be determined through calibration. The zoom parameter is calibrated by stepping through the allowable values and measuring the field of view after the motorized zoom is complete. User control of the pan/tilt head is used to actively and directly measure the field of view at each setting. Some visible landmark is chosen in the scene, roughly level to the pan/tilt device. The head is then directed by hand to find the left and right pan angles that bring the landmark to the far right and left edges of the image. Alternatively, an automated method based on self-calibration via active camera rotation can be used [8].

The relationship between focus parameter value and object distance is calibrated by focusing on objects at different distances from the camera, and deriving an implicit relationship between focus value and distance. This implicit relationship is represented as a lookup table of focus parameter settings, indexed by inverse distance to the desired focal distance in the scene. Focus to points at intermediate distances is determined by interpolation of these stored table values. A table indexed by inverse object distance is preferrable to one indexed by distance, since good results can be achieved using only linear interpolation on a sparse set of distance/focus measurements.

5. SAMPLE RESULTS

A three camera prototype system has been built in the Virtualized Reality lab at Carnegie Mellon University. This small demo system is used as follows. Initially, the room is empty, and a background model is acquired from each camera. A person then enters the room, and is detected by background subtraction and thresholding. Pixels that are determined to be part of the person's silhouette are used to estimate a normalized color histogram, which forms the appearance model for the mean-shift tracking algorithm. After the appearance model is acquired, the cameras begin active tracking and recording as the person moves throughout the space.

Figure 8-5 shows sample results from one recording session. Each row shows corresponding frames from each of the three cameras for a specific time sample. During recording, the demo tracking system automatically adjusted the pan and tilt parameters of each camera in real-time to keep the person's torso centered in each image.

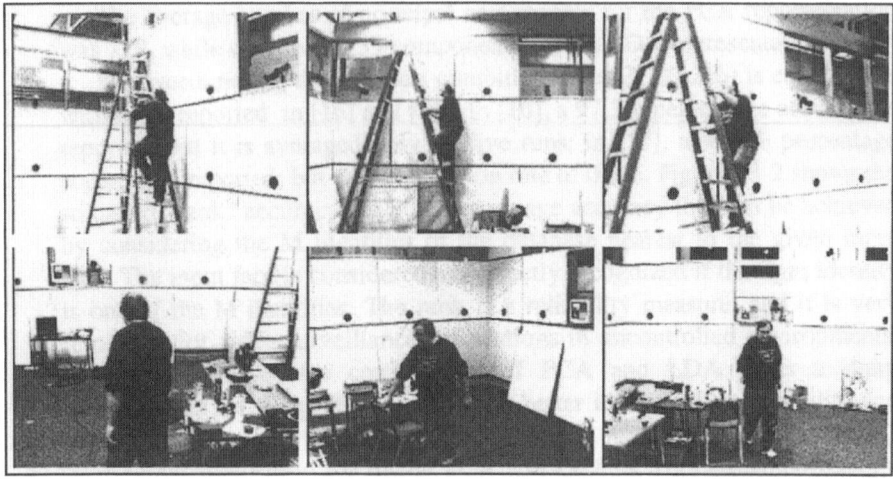

Figure 8-5. (Upper row) One time sample of synchronized multi-view video taken by a three-camera, real-time, active tracking system. (Lower row) Another time sample from the same recording session.

6. A MASTER-SLAVE VERSION FOR SPORTS BROADCASTING

The system described in previous sections is currently operating within a laboratory environment, where it is used for performing experiments in human activity recognition. The computer vision tracking algorithms work well in this controlled environment, but are not yet robust enough for commercial applications in natural outdoor environments. Motivated by a desire to push our technology into real-world applications, we have adapted our basic system design to produce a new system that uses a "human in the loop" to yield robustness in unconstrained environments.

The intended application area of the new system is sports broadcasting. Not only is it very difficult to automatically track a single player through a field full of other players, many dressed in similar uniforms, who are all bumping into and occluding each other, but it is far beyond current technology to automatically determine where the interesting events are occurring on the field. For this application, we have developed a multi-camera master-slave system that achieves real-time servoing on a moving point that a human cameraman can "steer" around the field. One of the goals of this robotic multi-camera system is the ability to freeze the action simultaneously from all cameras, so that playing one frame from each of the surrounding cameras sequentially gives the appearance of moving around the action while it is frozen in time. This heightens the viewer's ability to perceive the 3D spatial relationships between players, the ball, and field markers. A secondary goal is to greatly increase the number of viewing directions from which a play is simultaneously seen, so that there is a greater likelihood that a good viewpoint will be available to help referees adjudicate questionable plays, for example, to determine whether a player's foot went out of bounds, or whether the ball touched the ground before it was caught.

6.1 Master-Slave Adaptation

Assume there are N+1 cameras situated around a playing field. The cameraman remotely controls the pan, tilt, zoom and focus of one of the cameras (the master camera) to take shots of interesting events. For example, he may be keeping a runner centered in the field of view. In response to his pointing control, the system determines where on the field his desired servo point is, and automatically commands the remaining N cameras (slave cameras) to adjust pan, tilt, zoom and focus to keep the subject centered, properly sized, and focused in their fields of view (Figure 8-6).

During system operation, the cameraman can select any camera in the system to act as a master camera, and can change to a different master camera at any time. The cameraman controls the current master camera through a remote control gimbal (a Vinten Hawkeye unit was used in our system). The current master camera pan, tilt, zoom and focus parameters are read directly from this gimbal into a master computer, then relayed to the master camera platform.

Based on the pan/tilt angles of the master camera, the master computer determines the equation of a 3D line specifying the principle viewing ray of the master camera. All points on this line can be represented as $p = c + k\,v$, where p is a 3D point on the line, c is the focal point of the master camera, v is a unit vector representing the orientation of the principal axis, directed out from the focal point, and k is a scalar parameter that selects different points on the line. Only points on the line that are in front of the focal point $(k>0)$ are considered to be on the master camera principle viewing ray.

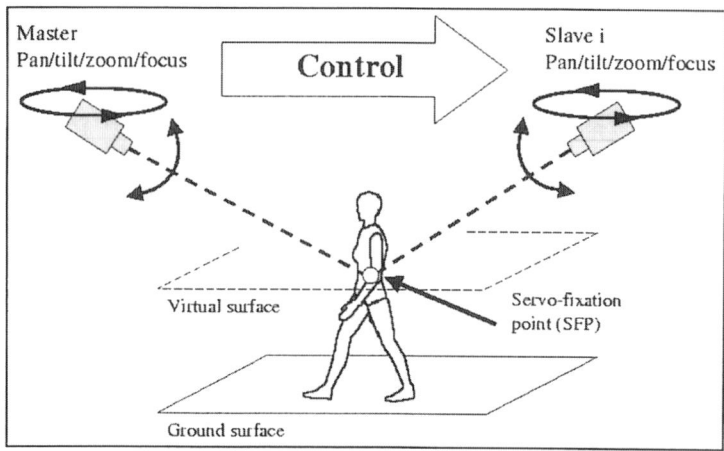

Figure 8-6. Master camera parameters are mapped onto a virtual surface to determine a 3D servo-fixation point within the dynamic scene. Each slave camera is automatically directed to view this point by computer control of its pan, tilt, zoom and focus parameters

There is of course an issue of determining the desired servo-fixation point (SFP) when only a single master camera is used to point at the action, since the SFP can occur anywhere along the principle viewing ray of the master camera. Choosing which point is the SFP is equivalent to choosing a value for parameter k in the above line equation describing the ray. In practice, we have had good results determining the SFP by intersecting the principle viewing ray with a virtual surface located a certain distance, e.g. four feet, above the playing field. This allows the camera operator to point at the torso of a player, rather than having to track the feet. After surveying a sparse set of points on the playing field, we represent the surface as a triangulated mesh, and elevate the surface to create the virtual surface for ray intersection by adding a constant value z_0 to the Z coordinates of each mesh point. For flat surfaces, mathematical intersection of the principle viewing ray with a plane equation would suffice, and would be very efficient. However, football fields, for example, are nonplanar in that they have a raised crest running down the length of the field to promote water runoff. The elevation of this crest can be as high as one or two feet above the elevation of the sidelines, and needs to be modeled for high accuracy pointing.

After determining the SFP, all computer-controlled slave cameras are tasked to achieve simultaneous alignment and focus at the target point in space, in a manner similar to the previously described system. The 3D SFP location and desired magnification factor are sent as commands from the master computer to each slave computer in UDP packets at a 30 Hz update rate. A software servo loop running on each slave camera computer is responsible for taking the sequence of computed goal parameters and controlling the slave cameras to smoothly and accurately track the field position designated by the master camera. To achieve smooth motion, each slave computer controls its pan/tilt head and camera/lens system at an even higher rate than it is receiving commands from the master camera. The slave computer interpolates between the last received command and the current command to control the slave camera pan, tilt, zoom and focus in smaller increments, more frequently.

Each camera is synchronized to a common genlock signal, so that the shutter for each camera fires at precisely the same time, resulting in video frames taken at the same time instant. Video frames from all cameras are then time-stamped, and stored digitally to enable fast retrieval of corresponding frames in time for all cameras. A user interface allows a human reviewer to then retrieve video frames temporally (sequential frames in time from a single camera) or spatially (the same time frame, retrieved from a sequence of cameras), assemble them into a new video sequence, and ultimately play the special effect over the network broadcast system.

6.2 The EyeVision System

In Fall 2000, we implemented this master-slave multi-camera servoing system in partnership with a major television network sponsor to create the EyeVision system. EyeVision was unveiled to the public on January 28, 2001, during the live broadcast of Superbowl XXXV in Tampa, Florida. Thirty camera heads were mounted on the upper deck of Raymond James Stadium, at intervals of roughly 7 degrees, to form a U-shape around one end of the field.

The spin image effect was shown several times during the game, and images from some of the 30 cameras were used to help the referees make a decision on one disputed play. Figure 8-7 shows sample views from a spin image sequence generated during the game. In all, the EyeVision experiment was successful, and will hopefully pave the way for future transfers of computer vision technology into the sports entertainment area, including systems with automated player detection and tracking algorithms, as well as more sophisticated real-time modeling and rendering capabilities.

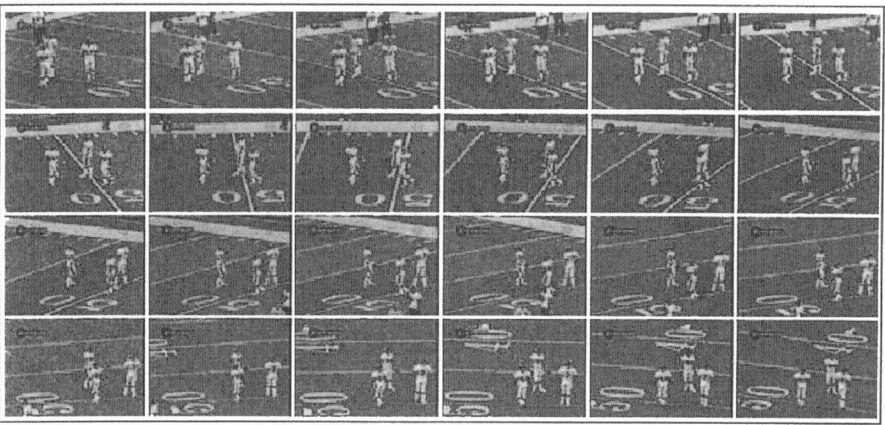

Figure 8-7. Example of spin-image views generated by the EyeVision system during the broadcast of Superbowl XXXV.

7. CONCLUSIONS

A system has been developed for acquiring multi-view video of a person moving through the scene. The approach uses a real-time appearance-based tracking algorithm to control the pan, tilt, zoom and focus parameters of multiple active cameras. The output of the system is a set of synchronized, time-stamped video streams of the person, seen simultaneously from several viewpoints. The system design emphasizes the modularity of each individual camera subsystem by minimizing the amount of information that each camera has about the other cameras. The cameras communicate by passing geometric "messages" through the shared 3D scene geometry, enabling a distributed approach to multi-camera active tracking.

A master-slave version of this system has also been developed. This adaptation allows a single human cameraman to control multiple active cameras to servo on a moving fixation point in the scene. The cameraman controls a master camera to follow the action on the field, and his desired fixation point in the scene is determined by intersecting the master camera principle viewing ray with a virtual surface four feet above the field. This 3D location is broadcast to the slave cameras, which adjust to point at that location, at an appropriate zoom and focus. This master-slave system has been successfully demonstrated in a commercial sports broadcasting application.

REFERENCES

[1] Collins, R., Lipton, A., Fujiyoshi, H. and Kanade, T., "Algorithms for cooperative multi-sensor surveillance," Proceedings of the IEEE, Vol. 89(10), pp.1456-1477, October 2001.

[2] Elgammal, A., Harwood, D. and Davis, L., "Non-parametric model for background subtraction," European Conference on Computer Vision, 2000.

[3] Stauffer, C. and Grimson, W.E.L., "Adaptive background mixture models for real-time tracking," IEEE Computer Vision and Pattern Recognition, 1999, pp.246-252.

[4] Toyama, K., Krumm, J., Brumitt, B. and Meyers, B., "Wallflower: Principles and practice of background maintenance," International Conference on Computer Vision, 1999, pp.255-261.

[5] Dellaert, F. and Collins, R., "Fast image-based tracking by selective pixel integration," ICCV Workshop on Frame-Rate Vision, September 1999.

[6] Bradski, G.R., "Computer vision face tracking for use in a perceptual user interface," IEEE Workshop on Applications of Computer Vision, 1998, pp. 214-219.

[7] Comaniciu, D., Ramesh, V. and Meer, P., "Real-time tracking of non-rigid objects using mean shift," IEEE Computer Vision and Pattern Recognition, 2000, pp.142-149.

[8] Collins, R. and Tsin, Y., "Calibration of an outdoor active camera system" IEEE Computer Vision and Pattern Recognition, June 1999, pp. 528-534.

Chapter 9

A COMPUTATIONAL FRAMEWORK FOR SIMULTANEOUS REAL-TIME HIGH-LEVEL VIDEO REPRESENTATION

Extraction of moving objects and related events

Aishy Amer
Concordia University
Electrical and Computer Engineering
Montréal, Québec, Canada

Keywords: Content-based video shot representation, video abstraction, video indexing, high-level content, semantic features, video objects, events, object extraction, video interpretation, video surveillance.

1. Introduction

Because of the ever-increasing needs for video content accessibility, developing automated and effective frameworks for *content-oriented* video representation have become an active field of research.

Typically, a video is a set of stories, scenes, and shots (Fig. 9.1). Examples of a video are a movie, a news clip, or a traffic surveillance clip. In movies, each scene is semantically connected to previous and following ones. In surveillance applications, a video does not necessarily have semantic flows.

A video contains, usually, thousands of shots. To facilitate automated content-oriented video representation, a video has to be first segmented into shots (see, e.g., [14, 7]). A shot is a (finite) sequence of images recorded contiguously (usually without viewpoint change) and represents a continuous, in time and space, action or event driven by moving objects (e.g., an intruder moving or an object stopping at a restricted site). There is little semantic change in the visual content of a shot, i.e., within a shot there is a short-term temporal consistency. (In the remainder of

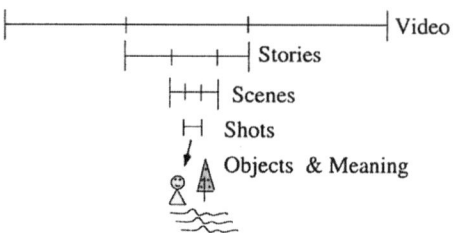

Figure 9.1. Video units.

this paper, the term video refers to a video shot.) A shot displays, in general, multiple *objects*, their *semantic interpretation* (i.e., objects' meaning), their *dynamics* (i.e., objects' movement, activities, action, or related events), and their *syntax* (i.e., the way objects are spatially and temporally related).

Developing effective video shot representation systems requires the resolution of two key issues: defining what are the most *important* and most *common* video contents and what *level* of features are suitable to represent these contents.

What are important video shot content? A video displays, in general, low and high-level features of objects within a given environment and context; an important observation is that the subject of the majority of video is related to moving objects, in particular people, that perform activities and interact creating object meaning such as events ([15, 6]).

What level of features are appropriate? When viewing a video, the human visual system (HVS) is, in general, attracted to moving objects and their features; the HVS focuses first on the high-level object features (e.g., meaning) and then on the low-level features (e.g., shape). The HVS is able to search a video by quickly scanning ("flipping") it for activities and interesting events. In addition, studies have shown that low-level features are not sufficient for effective video representation and that objects must be assigned high-level features as well ([6, 17]).

High-level intentional descriptions such as *what a person is thinking* can help solving video representation issues but extracting such information with the current state-of-the-art in video analysis is difficult and alternatives have to be found. Also extracting the context of a video data is difficult and may not be necessary to extract useful content as our investigation show. For example, a *deposit* event has a fixed semantic interpretation (an object is added to the scene) common to all applications but *deposit of an object* can have a variable meaning in different contexts.

9. A computational framework for high-level video representation

High-level object features are generally related to the movement of object and are divided into context independent and context dependent features. Context independent features include object movement, activity, action ([6]), and related events. High-level features are generally applicable when they convey fixed meaning independently of context.

To summarize, objects and high-level features are important and common for a wide range of advanced video applications, in particular to video surveillance applications. Context-independent features allow the design of widely applicable video representation systems.

High-level video representations can be structural and conceptual (Fig. 9.3 and 9.2). Structural representations are based on objects using spatial, temporal, and relational features while conceptual representations are based on movement-related features.

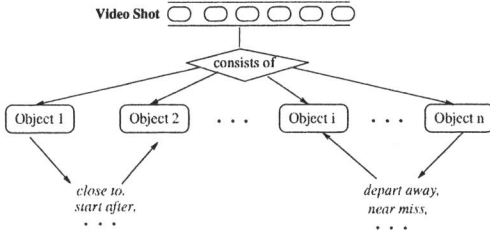

Figure 9.2. Structural video representation.

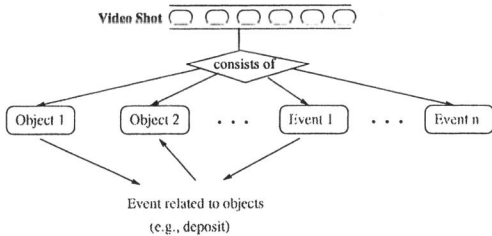

Figure 9.3. Conceptual video representation.

In this paper, we present a computational framework to conceptually represent an input video shot in real time based on its moving objects and related semantic features independent of context. Such high-level representation aims at assisting users of advanced applications, in particular video surveillance. To effectively represent video, our framework consist of four video processing layers (cf. Fig. 9.4): video enhancement to reduce noise and artifacts, video stabilization to compensate for global image changes, video analysis to extract low-level video features, and video interpretation to describe content in semantic-related terms.

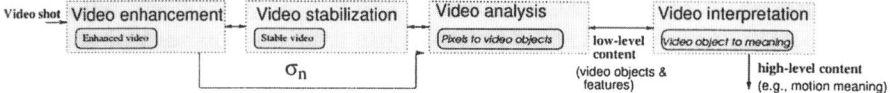

Figure 9.4. Our computational framework for video representation.

The paper is organized into six additional sections. Section 2 discusses related work, Section 3 describes our real-time framework for high-level video representation, Section 4 presents our method for analyzing video content to extract video objects and their low-level features, Section 5 presents our sub-framework for interpreting low-level spatio-temporal object features to extract context-independent semantic features, Section 6 presents experimental results, and Section 7 summarizes this paper.

2. Related work

Research in the area of detecting, tracking, and identifying people and objects has become a central topic in video processing ([6, 13, 15]). In recent years, research interest in content-based video representation has shifted from motion and tracking towards detection and recognition of activities, actions and events. Much work on video representation deals with the development of a generally applicable solution; however, there are few tests in the presence of noise and other artifacts. Most representation frameworks use mainly low-level features. Most high-level video representation frameworks are developed for narrow applications ([6]) and little work on context-*independent* representation exist.

Two basic video processing levels are required to represent high-level video content: analysis level to extract low-level content and interpretation level to describe content in semantic-related terms. Our framework includes these two levels and in the following we review related work.

Video analysis

The VideoQ system ([8]) uses video analysis based on optical flow, color, and edge features. Such a system has difficulties in the presence of large motion and occlusion. The AVI system ([9]) is based on motion detection using a background image and tracking using prediction and nearest-neighbor matching. The motion detection used is sensitive to noise and artifacts. The system is limited to indoor applications and cannot deal with occlusion. Recently, a new video analysis scheme, the COST-AM, has been introduced ([2]). This method is based on motion detection and color segmentation and gives good object masks. Difficulties arise when the combination of motion and color fails and strong artifacts are introduced (see Sec. 6). In addition, the method tends to lose objects which is critical for subsequent processing.

Video interpretation

Most high-level video representation are developed for narrow applications. Narrow-domain systems recognize events and actions, for example, in hand sign applications or in Smart-Cameras based cooking (see the special section in [18]), ([13, 20, 6]). In these systems, prior knowledge is, usually, inserted in the event recognition inference system and the focus is on recognition and logical formulation of events and actions.

Little work on context-*independent* or end-to-end video representation exist. The system in [9] is based on motion detection and tracking using prediction and nearest-neighbor matching. The system is able to detect basic events such as *deposit*. It can operate in simple environments where one human is tracked and translational motion is assumed. It is limited to applications of indoor environments, cannot deal with occlusion, and is noise sensitive. Moreover, the definition of events is not widely applicable.

The event detection system for indoor surveillance applications in [19] consists of object extraction and event detection modules. The event detection module classifies objects using a neural network. The classification includes: *abandoned object*, *person*, and *object*. The system is limited to one *abandoned object* event in unattended environments. The definition of *abandoned object* is specific to a given application. The system cannot associate abandoned objects and the person who deposited them.

The system in [11] is designed for off-line processing applications and uses domain knowledge to facilitate extraction of events (wildlife hunt events). The system in [12] tracks several people simultaneously and uses appearance-based models to identify people. It determines whether a person is carrying an object and can segment the object from the person. It also tracks body parts such as head or hands. The system imposes, however, restrictions on the object movements. Objects are assumed to move upright and with little occlusion. Moreover, it can only detect a limited set of events.

3. Overview of our computational framework

The presented end-to-end system is oriented to three requirements: 1) flexible object representations that are easily cooperatively searched for video applications such as surveillance, abstraction, indexing, and manipulation, 2) reliable, stable processing of video that foregoes the need for precision, and 3) low computational cost. The presented framework is designed to balance demands for effectiveness (solution quality) and efficiency (computational cost). Without real-time consideration, a representation can lose its applicability. On the other hand, framework stability is important for successful use.

The objective of this paper is to develop a *low-complexity automatic* framework for *stable* representation of video shots of real environments such as those with occlusions and coding artifacts. To achieve these requirements, our multi-layered framework 1) is divided into simple but effective tasks avoiding complex operations, 2) takes video noise level into account, and 3) corrects or compensates for estimation errors at the various processing steps. This framework involves four interacting processing layers: video enhancement (including noise estimation as in [5] and noise reduction as in [3]), video stabilization, video analysis (Sec. 4), and video interpretation (Sec. 5). The input to the video enhancement module is the original video and its output is an enhanced version of it. This enhanced video is then processed by the video analysis module which outputs low-level descriptions of the enhanced video. The video interpretation module takes these low-level descriptions and produces high-level descriptions of the original video.

Fig. 9.5 displays a block diagram of the presented framework where $R(n)$ represents a background image of the shot and σ_n is the estimated noise standard deviation. (Implemented modules are underlaid with gray boxes.) The presented framework can be viewed as a framework of methods and algorithms to build automatic dynamic scene interpretation and representation. Such interpretation and representation can be used in various video applications. Besides applications such as video surveillance and retrieval, outputs of the presented framework can be used in a video understanding or a symbolic reasoning framework. Our implementation of the video analysis and interpretation layers needs on between 0.1107 and 0.3507 seconds to extract content of two images on a SUN-SPARC-5 360 MHz.

In our framework, high-level features are related to moving objects and their semantic features. Moving objects are represented using quantitative and qualitative low-level features. Semantic features are represented by generic motion-related high-level features such as events. Semantic features are defined by approximate but efficient world models. This is done by continually monitoring changes and behavior of low-level features of the scene's objects. When certain conditions are met, high-level semantic features such as events are detected. To achieve higher applicability, descriptions are extracted independently of the context of

9. A computational framework for high-level video representation

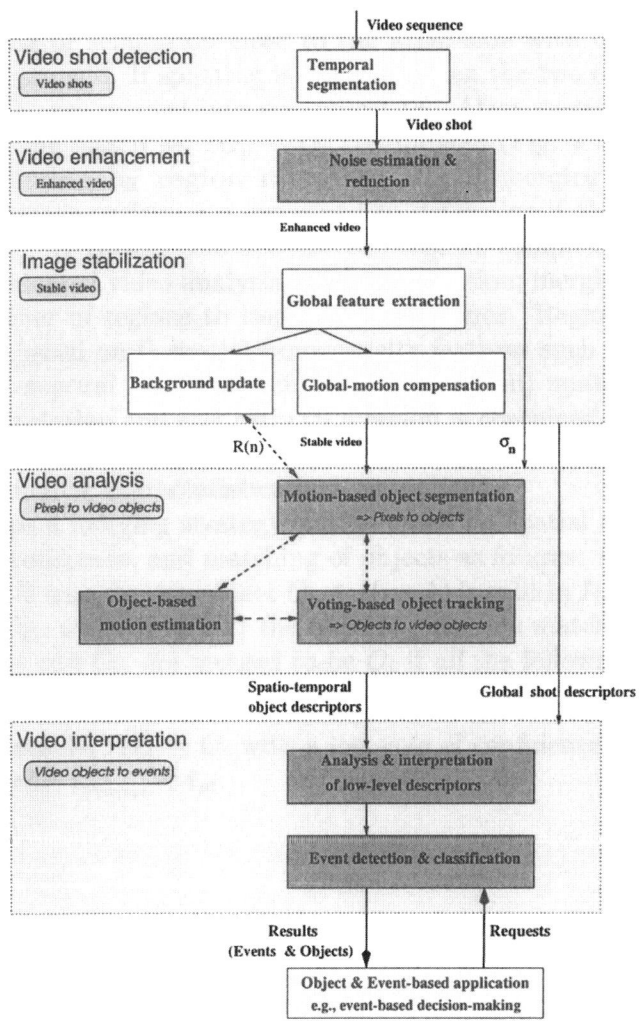

Figure 9.5. Presented framework for video shot representation.

the video. Several context independent events are rigorously defined and automatically detected using features extracted following segmentation, motion estimation and object tracking.

Our framework outputs at each instant n a list of objects with their *identity* throughout the shot, *low-level features* (location, shape, size, motion), *trajectory, life span or age, event descriptions*, and *spatio-temporal relationship*. This can be used in applications where an interpretation ("what is this video shot about?") is needed. For example, in video surveillance, event-oriented alarms can be activated.

A key contribution of our framework is the layer interaction for correction and compensation of processing errors at higher layers where more information is available. This includes 1) processing inaccuracies and false alarms– for example, the framework is able to differentiate between deposited objects, split objects, and objects at an obstacle and 2) data update– for example, low-level object segments are used for tracking and tracking can correct and merge these segments if needed; tracking together with merging are then used to detect events. Such interaction allow balanced processing as missing information can prevent complete information, while additional information may mask relevant information.

4. Object-oriented video analysis

Overview

The presented video analysis consists of: 1) motion-detection based object segmentation, 2) object-based motion estimation, 3) region merging, and 4) feature-voting based object tracking. Video analysis modules may produce inaccuracies and much research has been done to enhance their performance. We propose to compensate for errors of low-level steps at higher levels when more reliable information is available. The critical tasks of the video analysis are the segmentation and the tracking steps that are developed in the following sections.

Object segmentation

Object segmentation trades precise segmentation at object boundaries for speed of execution and reliability in varying image conditions. This interpretation is most appropriate to applications such as surveillance and video retrieval where speed and temporal reliability are of more concern than accurate object boundaries. The presented segmentation consists of four steps: motion detection, morphological edge detection, contour analysis, and object labeling. Motion detection provide binary images where foreground objects are marked white. It is adaptive as it uses estimated intra-image noise variance and detected inter-image motion. Edge detection find binary edge using new morphological operations with significantly reduced complexity compared to traditional morphological operations. Contour analysis transforms edges into contours and uses data from previous frames to adaptively eliminate unwanted contours. Object labeling transforms contours into objects using a Scan-Line contour filling method ([3]).

A memory-based motion detection method. The core of the segmentation method is the motion detection which must remain reliable throughout video shots. Motion can be detected either between successive images or between an image and a background image. A major difficulty with techniques using consecutive images is that they depend on inter-image motion being present between every image pair. A moving object (or part of an object) that becomes stationary or uncovered is erroneously merged with the background. Furthermore, temporal changes between two successive images may be detected in areas that do not belong to objects but close to object boundaries and in uncovered background. Also, removed or deposited objects are not detected.

A block diagram of our method to detect motion between the current image $I(n)$ and the background image $R(n)$ is shown in Fig. 9.6. The method comprises an averaging filter LP, a maximum filter max (Eq. 9.1), and spatio-temporal adaptation.

$$D(n) = max(LP(|I(n) - R(n)|)) \qquad (9.1)$$

In real images, the difference $|I(n) - R(n)|$ includes artifacts due to

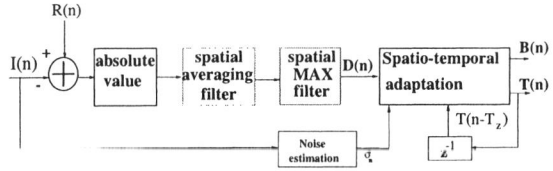

Figure 9.6. Proposed motion detection.

noise and illumination changes. To increase spatial accuracy of detection, an average and a maximum filter are used. Averaging causes a linear addition of the correlated true image data, whereas the noise is uncorrelated and is reduced by averaging. Hence, motion detection becomes less sensitive to noise and the difference image becomes smoother. The maximum operator limits motion detection to a neighborhood of the current pixel, causing stability around object boundaries and reducing granular noisy points.

Spatio-temporal adaptation. Typical difficulties of motion detection based on differencing are 1) it does not distinguish between object motion and other changes, for example, due to illumination changes and 2) it does not account for changes occurring throughout a long video shot. Usually a fixed threshold is used for all images of the shot to decide on moving and non-moving image parts. A fixed threshold method

fails, e.g., when the amount of moving regions changes significantly. To answer these difficulties, we propose a three step thresholding method.
1. **Adaptation to noise:** First, a spatial threshold, T_g, is estimated using a robust thresholding method [4]. The threshold T_g is adapted to the amount of image noise using the following positive-quadratic weighting:

$$T_n = T_g + c \cdot \sigma_n^2, \quad c < 1 \tag{9.2}$$

This weighting is a function of the noise standard deviation, σ_n, taking into consideration that low sensitivity to small σ_n is needed. σ_n is estimated using a new fast block-based noise estimation technique [3].
2. **Quantization:** To further stabilize thresholding, T_n is quantized to T_q into m values to compensate for background and local illumination changes. In our implementation, m was set to 3 and the quantization is:

$$T_q = \begin{cases} T_{\min} & : \ T_n \leq T_{\min} \\ T_{\mathrm{mid}} & : \ T_{\min} < T_n \leq T_{\mathrm{mid}} \\ T_{\max} & : \ \text{otherwise} \end{cases} \tag{9.3}$$

3. **Adding memory:** To adapt detection to temporal changes throughout a video shot the following memory function is used:

$$T(n) = \begin{cases} T_{\min} & : \ T_q \leq T_{\min} \\ T(n-1) & : \ T_q < T(n-1) \\ T_q & : \ \text{otherwise} \end{cases} \tag{9.4}$$

This function examines if there has been a significant change, i.e., $T_q > T(n-1)$, in the current image and, if so, the current threshold T_q is selected. If no significant temporal change is detected, i.e., $T_q < T(n-1)$, the previous threshold $T(n-1)$ is selected. When no or little motion is detected, $T_q \leq T_{\min}$, $T_{\min}$ is selected.

Simultaneous tracking of multiple objects

Objects are tracked based on the similarity of their features in successive images. This is done in three steps: object segmentation and motion estimation, object matching, and feature monitoring and correction (Fig. 9.7). In the first step, object segmentation and motion estimation segment and extract objects and computes their spatial and temporal features. The second, object matching, and third, feature monitoring and correction, steps are developed in the next sections.

In the second step, using a voting-based feature integration, *each* object O_p of the previous image $I(n-1)$ is matched with an object O_i of the current image $I(n)$ creating a unique correspondence $M_i : O_p \to O_i$. This means that all objects in $I(n-1)$ are matched with objects in $I(n)$.

9. A computational framework for high-level video representation

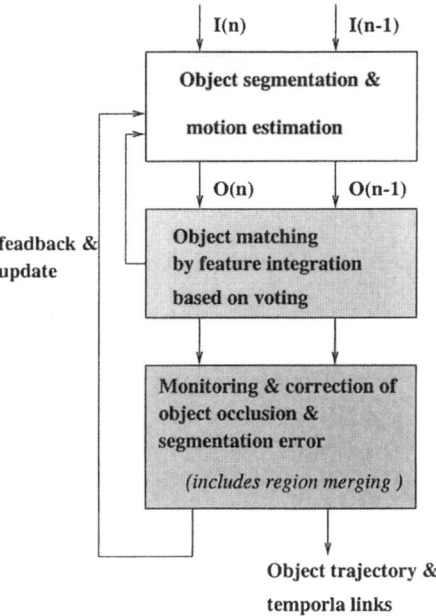

Figure 9.7. Framework of the presented tracking method.

In this step, each tracked object is assigned an *identity* throughout the image sequence. M_i provides a temporal link between objects to determine the trajectory of each object throughout the video and allows a semantic-based interpretation of the input video.

In the third step, feature monitoring and correction, object segmentation errors, such as when object occlude or are split, are detected and corrected. These new data are then used to *update* the results of previous steps, i.e., object segmentation and motion estimation (observe the feedback loops in Fig. 9.7). For example, the error correction steps can produce new objects after detecting occlusion. Motion estimation and tracking need to be performed for these new objects.

Step 2: Feature integration by voting. In this step, we combine spatial and temporal features using a non-linear voting scheme consisting of two steps: voting for object features of two objects (object voting) and voting for features of two correspondences in the case one object is matched to two objects (correspondence voting). Each voting step is first divided into m sub-votes with m object features. Since features can become harmful or occluded, the value m varies spatially (objects) and temporally (throughout the image sequence) depending on a spatial and temporal filtering. Then each sub-vote, m_i, is performed separately using an appropriate voting function. When the voting function is applied either a similarity variable s or a non-similarity variable d is increased. Depending on the number of features in a sub-vote, m_i, s or d may increase by one or more. Finally, a majority rule compares the two variables and decides about the final vote. The simplicity of the two-step non-linear feature combination which uses distinctive features based on properties of the HVS provides a good basis for fast and efficient matching which is illustrated in the results sections.

In case of zero match $\dashv O_i$, i.e., no object in $I(n-1)$ can be matched to an object in $I(n)$, a new object is declared entering or appearing into the scene depending on its location. In case of reverse zero match $O_p \dashv$, i.e., no object in $I(n)$ can be matched to an object in $I(n-1)$, O_p is declared disappearing or exiting the scene which depends on its location.

Object voting. In this step, three main feature votes are used: shape, size, and motion vote. The use of multiple votes aims at avoiding cases where one feature fails and the tracking module loses the object (especially in the case of occlusion). Define

- O_p, an object of the previous image $I(n-1)$;
- O_i, the ith object of the current image $I(n)$;
- $M_i : O_p \to O_i$, a correspondence and $\bar{M}_i = O_p \not\to O_i$ a non-correspondence between O_p and O_i;
- d_i, the distance between the centroids of O_p and O_i;
- t_r, the radius of a search area around O_p;
- $w_i = (w_{x_i}, w_{y_i})$, the estimated displacement of O_p relative to O_i;
- $w_{\max}$, the maximal possible object displacement (typically $15 < w_{\max} < 32$);
- s, the variable to count the similarity between O_p and O_i;
- d, the variable to count the dissimilarity between O_p and O_i;
- s_{++}, an increase of s by one vote; and
- d_{++}, an increase of d by one vote.

Then, where t_m is a threshold,

$$\begin{aligned} M_i &: \begin{array}{l} (d_i < t_r) \wedge (w_{x_i} < w_{y_{\max}}) \wedge \\ (w_{y_i} < w_{y_{\max}}) \wedge \\ (\zeta > t_m) \end{array} \\ \bar{M}_i &: \text{otherwise} \end{aligned} \qquad (9.5)$$

with the vote confidence $\zeta = \frac{s}{d}$. M_i is accepted if O_i lays within a search area of O_p, its displacements is not larger than a maximal displacement, and if both objects are similar, i.e., $\frac{s}{d} > t_m$. The use of this rule instead of the majority rule (i.e., $s > d$) is to allow the acceptance of M_i even if $s < d$. This is important in the case objects are occluded, where some features are significantly dissimilar and might cause the rejection of a good correspondence. Note that this step is followed by a correspondence step and no error can be introduced because of accepting correspondences with eventually dissimilar objects.

To compute the similarity variable s and the dissimilarity variable d between two objects O_i and O_j, three feature vote functions are applied. The three functions are based on the features shape, size, and motion (direction). Here we give the shape vote function as follows. Let $e_p(e_i)$, $c_p(c_i)$, $r_p(r_i)$ be the extent ratio, compactness and irregularity of the shape of $O_p(O_i)$, $d_{e_i} = |e_p - e_i|$, $d_{c_i} = |c_p - c_i|$, and $d_{r_i} = |r_p - r_i|$. Then

$$\begin{aligned} s_{++} &: d_{e_i} \leq t_s \quad \vee \quad d_{c_i} \leq t_s \quad \vee \quad d_{r_i} \leq t_s \\ d_{++} &: d_{e_i} > t_s \quad \vee \quad d_{c_i} > t_s \quad \vee \quad d_{r_i} > t_s \end{aligned} \quad (9.6)$$

Note that only if the two features significantly differ the vote is applied.

Correspondence voting. Recall that *all* objects $I(n-1)$ are matched against *all* objects of $I(n)$. Each object $O_p \in I(n-1)$ is matched to each object $O_i \in I(n)$. This may result in multiple matches for one object, for example, $(M_{pi} : O_p \to O_i$ and $M_{pj} : O_p \to O_j)$ or $(M_{pi} : O_p \to O_i$ and $M_{qi} : O_q \to O_i)$ with $O_p, O_q \in I(n-1)$ and $O_i, O_j \in I(n)$. If the final correspondence voting results in $s_i \approx s_j$, i.e, two objects of $I(n)$ are matched with the same object in $I(n-1)$, or $s_p \approx s_q$, i.e., two objects of $I(n-1)$ are matched with the same object in $I(n)$ plausibility rules are applied to resolve the ambiguity, as follows: Let s_i (s_j) be the variable that describes if M_i (M_j) is the better correspondence. Then

$$\begin{aligned} M_i &: s_i > s_j \\ M_j &: s_i \leq s_j \end{aligned} \quad (9.7)$$

A majority voting rule is applied here.

To compute the similarity variable s_i and the dissimilarity variable s_j between two correspondences, M_i and M_j five vote functions are applied. The five functions are based on the features distance, confidence, size, shape, and motion (direction and displacement). Here we give the motion vote function as follows.

- Direction vote: let $\delta_c = (\delta_{x_c}, \delta_{y_c})$, $\delta_p = (\delta_{x_p}, \delta_{y_p})$, $\delta_u = (\delta_{x_u}, \delta_{y_u})$ be the current, previous, and past-previous motion direction of O_p.

Let $\delta_i = (\delta_{x_i}, \delta_{y_i})$ be the motion direction of O_p if it is matched to O_i and $\delta_j = (\delta_{x_j}, \delta_{y_j})$ if matched to O_j.

$$\begin{aligned} s_{i++} : \quad & (\delta_{x_i} = \delta_{x_c} \wedge \delta_{x_i} = \delta_{x_p} \wedge \delta_{x_i} = \delta_{x_u}) \ \vee \\ & (\delta_{y_i} = \delta_{y_c} \wedge \delta_{y_i} = \delta_{y_p} \wedge \delta_{y_i} = \delta_{y_u}) \\ s_{j++} : \quad & (\delta_{x_j} = \delta_{x_c} \wedge \delta_{x_j} = \delta_{x_p} \wedge \delta_{x_j} = \delta_{x_u}) \ \vee \\ & (\delta_{y_j} = \delta_{y_c} \wedge \delta_{y_j} = \delta_{y_p} \wedge \delta_{y_j} = \delta_{y_u}) \end{aligned} \quad (9.8)$$

- Displacement vote: let d_{m_i} (d_{m_j}) be the displacement of O_p relative to O_i (O_j) and $d_m^k = |d_{m_i} - d_{m_j}|$. Then

$$\begin{aligned} s_{i++} &: \quad (d_m^k > t_m^k) \quad \wedge \quad (d_{m_i} < d_{m_j}) \\ s_{j++} &: \quad (d_m^k > t_m^k) \quad \wedge \quad (d_{m_i} > d_{m_j}) \end{aligned} \quad (9.9)$$

Step 3: Feature monitoring and correction. A good tracking technique must account for errors of previous steps. Object segmentation, for example, is likely to output erroneous object masks and features. Our method corrects or compensates effects of such errors. This is done based on plausibility rules and predictions strategies to filter faulty object features, to monitor occlusion, and to merge divided objects. The new information is then used to *update* the previous analysis steps (Fig. 9.7). The detection of segmentation errors is done based on an analysis of displacements of the four minimum bounding box (MBB) sides. This is shown in this paper by detection and correction of object occlusion and split.

Monitoring object occlusion. Let

- $O_{p_1}, O_{p_2} \in I(n-1)$,
- $M_i : O_{p_1} \to O_i$ where O_i results from the occlusion of O_{p_1} and O_{p_2} in $I(n)$,
- $d_{p_{12}}$ the distance of the centroids of O_{p_1} and of O_{p_2},
- $w = (w_x, w_y)$ the current displacement of O_{p_1}, i.e., between $I(n-2)$ and $I(n-1)$, and recall
- $d_{r_{\max}}$ ($d_{r_{\min}}$) is the vertical displacement of the lower (upper) row and
- $d_{c_{\max}}$ ($d_{c_{\min}}$) is the horizontal displacement of the right (left) column of O_{p_1}.

9. A computational framework for high-level video representation

Object occlusion is declared if

$$(((|w_y - d_{r_{\max}}| > t_1) \wedge (d_{r_{\max}} > 0) \wedge (d_{i_{12}} < t_2)) \vee \\ ((|w_y - d_{r_{\min}}| > t_1) \wedge (d_{r_{\min}} > 0) \wedge (d_{i_{12}} < t_2)) \vee \\ ((|w_x - d_{c_{\max}}| > t_1) \wedge (d_{c_{\max}} > 0) \wedge (d_{i_{12}} < t_2)) \vee \\ ((|w_x - d_{c_{\min}}| > t_1) \wedge (d_{c_{\min}} > 0) \wedge (d_{i_{12}} < t_2))) \tag{9.10}$$

where t_1 and t_2 are thresholds. If occlusion is detected then both the occluding and occluded objects are labeled with a special flag. This labeling enable our system to continue tracking both objects in following images even if they are completely non visible. Tracking non visible objects is important since they might reappear. The labeling is further important to help detect occlusion even if the occlusion conditions in Eq. 9.10 are not met.

Figure 9.8. Object occlusion: large outward displacement of a MBB-side.

Correction of occlusion by prediction: If occlusion is detected, the occluded object O_i is split into two objects. This is done by predicting both object O_{p_2} and O_{p_1} onto $I(n)$ using the following displacement estimate:

$$d_{p_1} = (\text{MED}(d^1_{x_c}, d^1_{x_p}, d^1_{u_x}), \text{MED}(d^1_{y_c}, d^1_{y_p}, d^1_{u_y})) \\ d_{p_2} = (\text{MED}(d^2_{x_c}, d^2_{x_p}, d^2_{u_x}), \text{MED}(d^2_{y_c}, d^2_{y_p}, d^2_{u_y})) \tag{9.11}$$

with MED represent a 3-tap median filter, $d^1_{x_c}(d^1_{y_c})$, $d^1_{x_p}(d^1_{y_p})$, $d^1_{u_x}(d^1_{u_y})$ as the current, previous and past-previous horizontal (vertical) displacement of O_{p_1} and $d^2_{x_c}(d^2_{y_c})$, $d^2_{x_p}(d^2_{y_p})$, $d^2_{u_x}(d^2_{u_y})$ as the current, previous and past-previous horizontal (vertical) displacement of O_{p_2}. After splitting occluded and occluding objects, the list of objects of $I(n)$ is *updated*, for example, by adding O_{p_2}. Then a feedback loop estimate the correspondences in case new objects are added (Fig. 9.7).

An example of object occlusion detection and correction is shown in Fig. 9.9. The scene shows two objects moving and then they occlude each other. The change detection module provides one segment for both objects but the tracking module is able to correct the error and track the two objects also during occlusion.

Figure 9.9. An example of tracking two objects during occlusion.

Monitoring object splitting. Assume $O_p \in I(n-1)$ is split in $I(n)$ into two objects O_{i_1} and O_{i_2} (see Fig. 9.10(a)). Let $M_i : O_p \to O_{i_1}$, $d_{i_{12}}$ be the distance between the centroids of O_{i_1} and of O_{i_2}, and $w = (w_x, w_y)$ the current displacement of O_p between $I(n-2)$ and $I(n-1)$. Then object splitting is declared if

$$\begin{aligned}(|w_y - d_{r_{\max}}| > t_1) \wedge d_{r_{\max}} < 0 \wedge d_{i_{12}} < t_2 \quad &\vee \\ (|w_y - d_{r_{\min}}| > t_1) \wedge d_{r_{\min}} < 0 \wedge d_{i_{12}} < t_2 \quad &\vee \\ (|w_x - d_{c_{\max}}| > t_1) \wedge d_{c_{\max}} < 0 \wedge d_{i_{12}} < t_2 \quad &\vee \\ (|w_x - d_{c_{\min}}| > t_1) \wedge d_{c_{\min}} < 0 \wedge d_{i_{12}} < t_2 \end{aligned} \quad (9.12)$$

This means if 1) the difference between the current object (O_1) displacement and the displacement of *one* of the four sides of the MBB is larger than a threshold, 2) the displacement of that MBB side is inwards (i.e., towards the center of the object), and 3) there is an object O_2 close to O_1 then object splitting or separation close to the MBB-side with the large displacement is assumed. If splitting is detected, then the two object regions O_{i_1} and O_{i_2} are merged into one object O_i. After merging the features of O_i and the match $M_i : O_p \to O_i$ are updated (Fig. 9.7).

Correction of splitting by region merging: Region merging is a process where contiguous regions are examined to determine if they can be merged [10, 16]. It is desirable because sub-regions complicate the process of object oriented video analysis and interpretation; merging reduces the total number of regions to improve performance. Regions can be merged either based on i) spatial homogeneity features such as texture or color, ii) temporal features such as motion, or iii) spatial relationships such as inclusion and size ratio (if a region is contained in another region and its size is significantly smaller, it maybe merged if the two objects show similar characteristics such as motion).

This section develops a merging strategy that is based on spatial relationships, temporal coherence, and matching of objects as follows: assume 1) equation 9.12 is true, i.e., an object $O_p \in I(n-1)$ is split in $I(n)$ into two sub-regions O_{i1}, and O_{i2} and 2) the matching process matches O_p with O_{i1}. Then O_{i2} and O_{i1} are merged to be O_i if *all* the following conditions are met:

- Object voting gives $M_i : O_p \to O_i$ with a low vote of confidence ζ, i.e., $\zeta > t_{m_{\text{merge}}}$ with $t_{m_{\text{merge}}} < t_m$.

9. A computational framework for high-level video representation

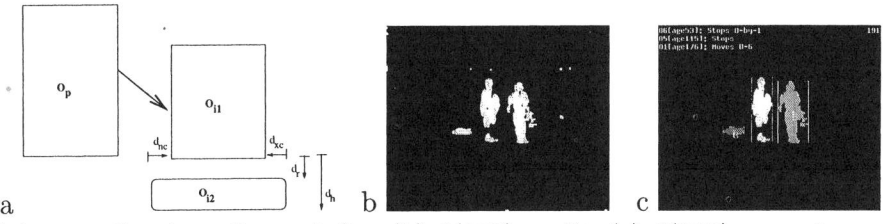

(a) spatially close O_{i1} and O_{i2}; (b) $I(190)$: split; (c) $I(190)$: merging.

Figure 9.10. Region merging examples.

- If a split is found on one side of the MBB (based on Eq. 9.12), then all the displacements of the three other MBB sides of O_p should not change significantly when the two objects are merged.
- O_{i1} is spatially close to O_{i2} and O_{i2} to O_p, for example, in the case of down split as shown in Fig. 9.10(a), all the distances d, d_{nc}, d_{xc}, and d_{xr} are small.
- The size, hight, and width, of the merged object $O_i = O_{i1} + O_{i2}$ matches those of O_p. For example, $t_{\min} < \frac{A_p}{A_i} < t_{\max}$, with thresholds $t_{\min}, t_{\max}$.
- The motion direction of O_p does not significantly change if matched to O_i.

This merging strategy has proven to be powerful in various simulations. The good performance is due to the cooperation between the matching and merging processes. Each process supports the other based on rules that aim at limiting erroneous merging. The advantage of the proposed merging strategy compared to known merging techniques (cf. [10, 16]) is that it is based on temporal coherence throughout the tracking process. Fig. 9.10 shows an example of the performance of the merging process. As shown, the method is successful also when multiple small objects are close to the split object.

5. Context-independent video interpretation

Overview

Content-oriented video applications such as surveillance, require the development of automatic and real-time systems to extract high-level video features. In this section, we propose a video interpretation module (Fig. 9.5) that defines and extract semantic features based on approximate but efficient world models. We propose perceptual descriptions of semantic feature that are common for a wide range of applications.

Semantic feature detection is not based on geometry of objects but on their features and relations over time. This is done by continually monitoring changes and behavior of low-level features of the scene's objects. When certain conditions are met, high-level semantic features such as events are detected. Our context-independent interpretation module is applied here to detects events related to moving objects.

An event expresses a particular behavior of a finite set of objects in a sequence of a small number of consecutive images of a video shot. An event consists of context-dependent and context-independent (or fixed meaning) components associated with a time and location. For example, a *deposit* event has a fixed semantic interpretation (an object is added to the scene) common to all applications but *deposit of an object* can have a variable meaning in different contexts. In the simplest case, an event is the appearance of a new object into the scene or the exit of an object from the scene. In more complex cases, an event starts when the behavior of objects changes. An event detection technique should automatically and efficiently provide generally useful events.

The video analysis module represents objects by their spatio-temporal low-level features in temporally linked lists. Object features are stored for each image and compared as images of a shot arrive. To detect events, the video interpretation module simultaneously monitors the behavior and features of each object in the scene. Events are automatically extracted in a straightforward intuitive manner by combining trajectory information with spatial features, such as size and location. When specific conditions are met, events related to these conditions are detected. Behavior monitoring is done on-line, i.e., object data is analyzed *as it arrives* and events are detected as they occur.

The following are samples of events detected automatically by our framework (due to space constraints only selected events are defined here). The thresholds used in the following rules are adapted to object features. For example, the threshold, $t_{d_{\min}}$, when detecting *abnormal movements* is a function of the frame-rate, the motion, and the image size. Some thresholds are computed experimentally. However, the same values were taken for all shots used in simulations. In the following, let $I(n)$ be an image in a video shot and O_i a segmented object in $I(n)$.

Basic events

Basic events are related to trivial behavior of objects. Examples are *enter*, *appear*, *exit*, *disappear*, *move*, and *stops*. Following are definitions of some of the basic events our system is able to detect.

9. A computational framework for high-level video representation

Move. An object, $O_i \in I(n)$, *moves* in image $I(n)$ if 1) $M_i : O_p \to O_i$ (a function assigning O_p at time $n-1$ to an object O_i at time n) where $O_p \in I(n-1)$, and 2) the median of the motion magnitudes of O_i in the last k images is larger than a threshold.

Disappear. An object, O_p, *disappears* at time n if 1) $O_p \in I(n-1)$, 2) $O_p \notin I(n)$, i.e., zero match $M_0 : O_p \dashv$, 3) c_p is *not* at the border in $I(n-1)$, and 4) $g_p > t_g$.

Intra-object events

An intra-object event is related to non-trivial behavior of an object. Examples are *abnormal movements* and *dominant movements*. An abnormal movement is defined when, for example, an object *stays long* or *moves (too) fast/slow*. A dominant movement is given when an object, for examples, 1) performs a significant event, 2) has the largest size, 3) has the largest speed, 4) or has the largest age. Following are definitions of some of the intra-object events our system is able to detect.

Stays long. An object, O_i, *stays long* in the scene if 1) $g_i > t_{g_{\max}}$, i.e., O_i does not leave the scene after a given time. $t_{g_{\max}}$ is a function of the frame-rate and the minimal allowable speed, and 2) $d_i < t_{d_{\min}}$, i.e, the distance, d_i, between the current position of O_i in $I(n)$ and its past position in $I(l)$, with $l < n$ is less than a threshold $t_{d_{\min}}$ which is a function of the frame-rate, the motion, and the image size.

Dominant object. A *dominant object* 1) is related to a significant event, 2) has the largest size of all objects, 3) has the largest speed, or 4) has the largest age.

Inter-object events

An inter-object event is related to non-trivial behavior of two or more objects. Examples include objects *at an obstacle*, *occlude/occluded*, *expose/exposed*, *deposit/deposited*, and *remove/removed*. Following are definitions of some of these events.

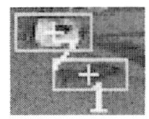

Occlusion. Object occlusion is declared as defined in Eq. 9.10 (see Page 162). With occlusion, at least two objects are involved where one is moving. With two objects, the object with the larger area is defined as the *occluding*

object, the other the *occluded object*. Note that *exposure* is detected when occlusion ends.

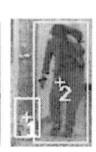

Removal.
Let $O_i \in I(n)$ and $O_p, O_q \in I(n-1)$ with $M_i : O_p \to O_i$. O_p removes O_q if

- O_p and O_q were occluded in $I(n-1)$,
- $O_q \notin I(n)$, i.e., zero match $M_0 : O_q \dashv$, and
- the area A_q of O_q is smaller than that of O_i, i.e., $\frac{A_q}{A_i} < t_a$, $t_a < 1$ being a threshold.

Removal is detected after occlusion. When occlusion is detected the tracking technique predicts the occluded objects. In case of removal, the features of the removed object can change significantly and the tracking framework may not be able to track the removed objects. Conditions for removal are checked and if they are met, removal is declared. The object with the larger area is the remover, the other is the removed object.

Deposit.
O_i deposits O_j (or O_j is *deposited* by O_i) if

- $O_p \in I(n-1)$, $O_i, O_j \in I(n)$, and M_i,
- $O_j \notin I(n-1)$, i.e., no match $M_0 :\dashv O_j$,
- $\frac{A_j}{A_i} < t_a$, $t_a < 1$ being a threshold,
- $A_i + A_j \simeq A_p \;\wedge\; [(H_i + H_j \simeq H_p) \vee (W_i + W_j \simeq W_p)]$, where A_i, H_i, and W_i are area, height, and width of O_i,
- O_j is close to a side, s, of the minimum bounding box (MBB) of O_i. $s \in \{r_{\min_i}, r_{\max_i}, c_{\min_i}, c_{\max_i}\}$. Let d_{is} be the distance between the MBB-side s and O_j. O_j is close to s if $t_{c_{\min}} < d_{is} < t_{c_{\max}}$ with thresholds $t_{c_{\min}}$ and $t_{c_{\max}}$, and
- O_i changes in height or width between $I(n-1)$ and $I(n)$ at the MBB-side s.

Note that only if the distance between the deposited object and depositor is large is the event *deposit* considered (in the real world, a depositor moves away from the deposited object). Otherwise O_j is assumed have split from O_i and is merged to O_i. To reduce false alarms, *deposit* is declared if the deposited object remains in the scene for some time. Thus the framework is able to differentiate between deposit and segmentation errors (e.g., object split). It can also differentiate between stopping objects (e.g., seated person or stopped car) and deposited objects.

Objects at an obstacle.
Often, objects move close to static background objects (called obstacles) that can occlude part of the moving objects. This is particularly frequent in traffic scenes where objects move close to traffic and other road signs. In this case, a change detection module is not able to detect pixels occluded by the obstacle and objects are split into two or more objects as shown in these figures:  ⇒ . This is different from object split because no abrupt, but a gradual, change of object size and shape occurs. This section develops a method to detect the motion of objects at obstacles. The method monitors the size of each object, O_i, in the scene. If a *continuous* decrease or increase of the size of O_i is detected (by comparing the area of two corresponding objects), a flag for O_i is set accordingly. Let $O_q, O_p \in I(n-1)$. Then O_q is at *an obstacle* if the following conditions are met:

- O_q and O_p have appeared or entered,
- O_q has no corresponding object in $I(n)$, i.e., zero match in $I(n)$ with $M_0 : O_q \dashv$,
- A_q was monotonically decreasing in the last k images,
- O_q has a close object O_p where
  - A_p was continuously increasing in the last k images and
  - O_p has a corresponding object, i.e., $M_i : O_p \rightarrow O_i$, with $O_i \in I(n)$.
- O_q and O_p have some similarity, i.e., object voting (Eq. 9.5) gives $M_p : O_q \rightarrow O_p \rightarrow O_i$ with a low confidence, and

- motion direction of O_q does not change if matched to O_i.

Note that while the transition images show two objects, the original object gets its ID back when motion at the obstacle is detected.

Other extensions

The above events are sufficiently broad for a wide range of applications to assist on-line supervision of, for example, the removal/deposit of objects in a surveillance site and the behavior of traffic objects. Other events can be easily extracted based on our interpretation strategy. We propose here the following extensions.

Composite events. Examples are: O_i *moved, stopped, is occluded,* and *reverses directions.* O_j *is exposed, moves,* and *exits.*

Stand/Sit/Walk. *Standing* and *sitting* are characterized by continuous change in height and width of the object MBB. *Sitting* is characterized by continual increase of the width and decrease of the height. When an object stands, the width of its MBB continual increase while the height decrease. In both events, height and width must be compared to the values of the height and width at the time instant before they started to change. The event *walk* can be easily detected as continuous moderate movements of a person.

Approaching a restricted site. This is an event that is straightforward to detect. If the location of a restricted site is given, the direction of an object's motion and distance to the site can be monitored and the event *approach a restricted site* can be eventually declared.

Object lost/found. At a time instant n, an object is declared lost if it has no corresponding object in the current image and occlusion was previously reported (but no removal). It is similar to the event *disappear.* Some applications require the search for lost objects even if they are not in the scene. To allow the framework to find lost objects, features, such as ID, size, or motion, of lost objects need to be stored for future reference. If the event *object lost* was detected and a new object appears in the scene which shows similar features to the lost object, the objects can be matched and the event *object found* declared.

Changing direction or speed. Based on a registered motion direction, which is registered when the object completely enters the scene, the motion direction in the last k images previous to $I(n)$ are compared with the registered motion direction. If the current motion direction is deviating from the motion direction in each of the k images, a change of direction can be declared. Similarly, change of speed can be detected.

Normal behavior. Often, a scene contains events which have never occurred before or occur rarely. This is application dependent. In general, normal behavior can be defined as a chain of simple events: for example, enters, moves through the scene, and disappears.

Object history. For some video applications, a summary or a detailed description of the spatio-temporal object features is needed. The proposed framework can provide such a summary. An object history can include: initial location, trajectory, direction and velocity, significant change in speed, spatial relation to other objects, distance between current location of the object and a previous location.

6. Results and applications

To test the performance of our framework, we conducted extensive experiments on shots containing over 6000 images with multi-object occlusion, noise, and artifacts of indoor and outdoor real environments. The presented framework works in real time for video shots with a rate of up to 10 frames/second. In this section, we first introduce sample results of the presented video analysis module and then show how the video interpretation module can be used for different applications.

Results of the video analysis module

The current implementation of the presented video analysis requires on average between 0.11 and 0.35 seconds to analyze the content of two images on a SUN-SPARC-5 360 MHz. For comparison, the current version (v.4x) of the reference method, COST-AM ([2]), takes on average 175 seconds to segment objects of an image.

Samples of our results are shown in this section. Both objective and subjective evaluations, and comparisons to other methods show the robustness of the presented methods while being of reduced complexity. For example, our motion detection method is compared to the detection method in [22] which uses a background images and a statistical motion detection method ([1]). The presented method displays better reliability especially in images with local illumination change and noise (cf. Fig. 9.11 where the reference method is [1] and [22]). In addition, the presented method has lower computational cost than the reference method.

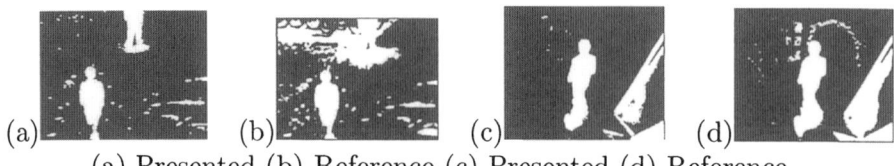
(a) Presented (b) Reference (c) Presented (d) Reference

Figure 9.11. Motion detection comparison.

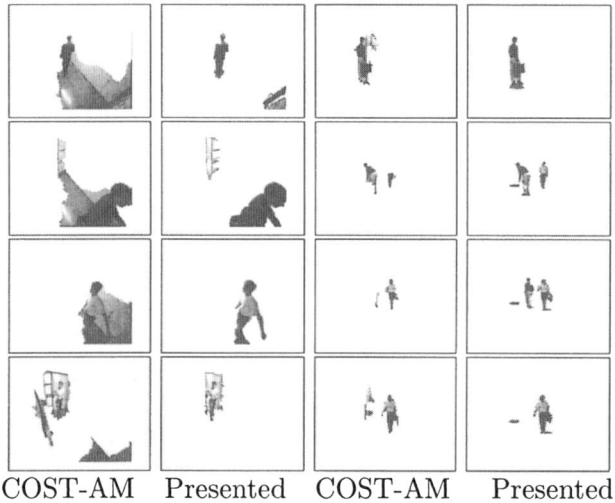

COST-AM　　Presented　　COST-AM　　Presented

Figure 9.12. Subjective comparison of object segmentation for the 'Stair' (left) and 'Hall' (right) shots.

Furthermore, the presented segmentation is objective and subjectively compared to the current version (4.x) of the COST-AM method ([2]). For example, Fig. 9.13 gives objective comparison results based on the criteria given in [21]. The presented segmentation is better with respect to all three criteria, especially it yields higher spatial accuracy. Fig. 9.12 subjectively confirms the good performance of the presented segmentation compared to the reference, COST-AM, method. COST-Am method loses some objects and its spatial accuracy is poor. The presented method remains robust to variable object size and is spatially more accurate. Robustness is further confirmed in the presence of MPEG-2 artifacts (25dB) and noise (30dB). The presented segmentation algorithm needs a maximum of 0.15 seconds on a SUN-SPARC-5. Fig. 9.14 and 9.15 display reliably estimated trajectories using the presented tracking method.

9. A computational framework for high-level video representation

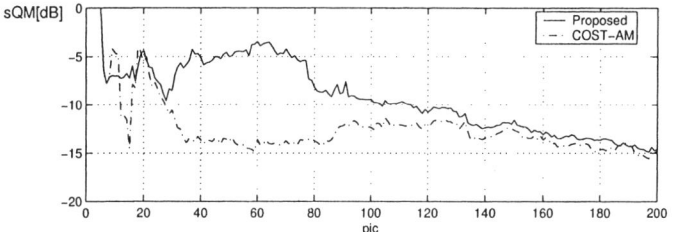

(a) Spatial accuracy, $sQM(dB)$, comparison: our method has better accuracy throughout the shot. Average gain $\simeq$ 3.5 dB

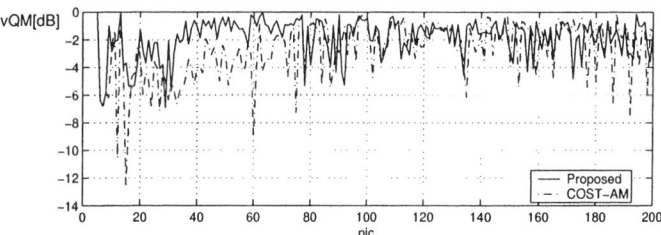

(b) Temporal stability, $vQM(dB)$, comparison: our method has higher stability throughout the shot. Average gain $\simeq$ 1.0 dB

Figure 9.13. Objective comparison of object masks for the 'Hall' shot.

Figure 9.14. Estimated trajectories of the objects in the shot 'Highway' using the presented video analysis framework. 'StartP' is start position.

Figure 9.15. Estimated trajectories of the objects in the shot 'Survey' using the presented video analysis framework. 'StartP' is start position.

Applications of the video interpretation module

Applications of our video representation framework include key-image extraction based on events, high-level video abstraction and summarization, and event-related alerts for surveillance applications.

Event-based video summary. The performance of our video interpretation is illustrated here by the following summaries of the shots 'floor' (826 images) and 'stair' (1475 images).

```
'floor' Shot Summary based on Objects and Events; StartPic 1/EndPic 826
---------------------------------------------------------------------------
|Pic| Event              | Obj| Age| Status  | Position           | Motion      | Size           |
|   |                    |    |    |         | start/present      | present     | start/present  |
|---|--------------------|----|----|---------|--------------------|-------------|----------------|
|36 | Appear             | 1  | 8  | Move    | (126,140)/(123,135)| (0  ,-1 )  | 320    /320    |
|---|--------------------|----|----|---------|--------------------|-------------|----------------|
|268| is Deposit by Obj 1| 3  | 8  | Stop    | (121,140)/(121,140)| (0  ,0  )  | 539    /539    |
|---|--------------------|----|----|---------|--------------------|-------------|----------------|
|405| Occlusion          | 3  | 145| Stop    | (121,140)/(120,141)| (0  ,0  )  | 541    /555    |
|---|--------------------|----|----|---------|--------------------|-------------|----------------|
|405| Occlusion Obj 3    | 1  | 377| Move    | (126,140)/(83 ,109)| (1  ,0  )  | 840    /2422   |
|---|--------------------|----|----|---------|--------------------|-------------|----------------|
|411| Removal by Obj 1   | 3  | 150| Removal | (121,140)/(104,132)| (0  ,0  )  | 541    /1451   |
|---|--------------------|----|----|---------|--------------------|-------------|----------------|
|787| Appear             | 18 | 8  | Move    | (105,68 )/(108,86 )| (0  ,0  )  | 91     /91     |
|---|--------------------|----|----|---------|--------------------|-------------|----------------|
|825| Exit               | 1  | 796| Exit    | (126,140)/(9  ,230)| (-10,7  )  | 840    /247    |
|---|--------------------|----|----|---------|--------------------|-------------|----------------|
```

```
'stair_wide_cif' Shot Summary based on Objects and Events;; StartPic 1/EndPic 1475
---------------------------------------------------------------------------
|Pic | Event     | Obj| Age| Status    | Position           | Motion      | Size           |
|    |           |    |    |           | start/present      | present     | start/present  |
|--- |-----------|----|----|-----------|--------------------|-------------|----------------|
|172 | Enter     | 2  | 8  | Move      | (312,248)/(308,230)| (-2 ,-2 )  | 5746   /5746   |
|--- |-----------|----|----|-----------|--------------------|-------------|----------------|
|216 | Enter     | 3  | 8  | Move      | (184,186)/(167,169)| (-2 ,-2 )  | 7587   /7587   |
|--- |-----------|----|----|-----------|--------------------|-------------|----------------|
|234 | Exit      | 2  | 69 | Exit      | (312,248)/(337,282)| (3  ,16 )  | 11803  /180    |
|--- |-----------|----|----|-----------|--------------------|-------------|----------------|
|479 | Appear    | 4  | 8  | Stop      | (128,104)/(127,100)| (0  ,0  )  | 211    /211    |
|--- |-----------|----|----|-----------|--------------------|-------------|----------------|
|547 | Disappear | 3  | 338| Disappear | (184,186)/(114,67 )| (0  ,-1 )  | 6374   /137    |
|--- |-----------|----|----|-----------|--------------------|-------------|----------------|
|608 | Enter     | 7  | 8  | Move      | (120,88 )/(125,79 )| (1  ,1  )  | 2536   /2536   |
|--- |-----------|----|----|-----------|--------------------|-------------|----------------|
|678 | Enter     | 8  | 8  | Move      | (138,85 )/(131,85 )| (-1 ,0  )  | 5308   /5308   |
|--- |-----------|----|----|-----------|--------------------|-------------|----------------|
|803 | Exit      | 7  | 202| Exit      | (120,88 )/(337,282)| (3  ,20 )  | 3432   /199    |
|--- |-----------|----|----|-----------|--------------------|-------------|----------------|
|812 | Disappear | 8  | 141| Disappear | (138,85 )/(121,92 )| (0  ,0  )  | 4334   /73     |
|--- |-----------|----|----|-----------|--------------------|-------------|----------------|
|916 | Enter     | 9  | 8  | Move      | (11 ,151)/(23 ,159)| (3  ,0  )  | 2866   /2866   |
|--- |-----------|----|----|-----------|--------------------|-------------|----------------|
|1179| Exit      | 9  | 270| Exit      | (11 ,151)/(127,72 )| (0  ,0  )  | 4955   /1388   |
|--- |-----------|----|----|-----------|--------------------|-------------|----------------|
|1290| Enter     | 16 | 8  | Stop      | (123,72 )/(128,77 )| (0  ,0  )  | 2737   /2737   |
|--- |-----------|----|----|-----------|--------------------|-------------|----------------|
|1363| Exit      | 16 | 80 | Exit      | (123,72 )/(83 ,154)| (-5 ,1  )  | 3844   /15910  |
|--- |-----------|----|----|-----------|--------------------|-------------|----------------|
```

Event-based key-image extraction. In a surveillance environment, important events may occur after a long time has passed. During this time, the attention of human operators decreases and significant events may be missed. The proposed system for event detection identifies events of interest as they occur and human operators can focus their attention on moving objects and their related events.

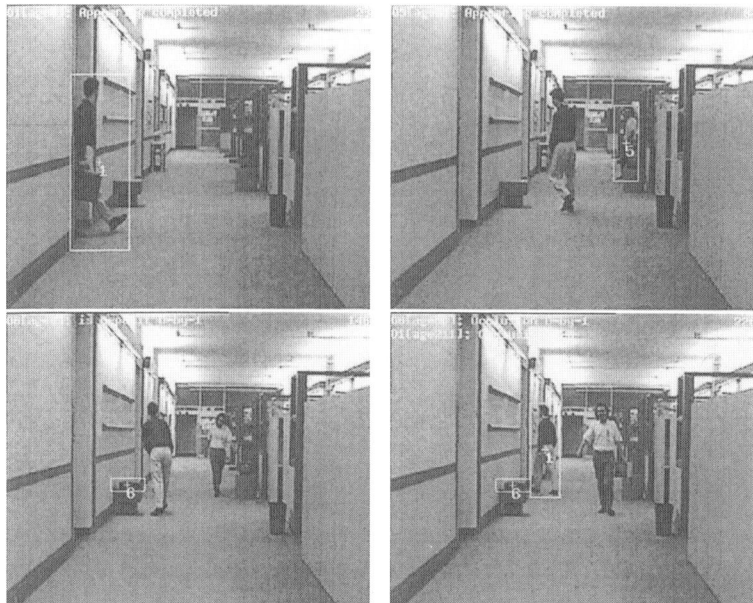

Figure 9.16. Key events of the 300 images of the 'Hall' shot: O_6 is deposited by object O_1.

This section presents automatic extracted key-images from video shots using our framework. Key-images are the subset of images which best represent the content of a video sequence in an abstract manner. Key-image video abstraction transforms an entire video shot into a small number of representative images. This way important content is maintained while redundancies are removed. Key-images based on events are appropriate when the system must report on specific events as soon as they happen.

Figs. 9.16, 9.17, and 9.18 show images of key events extracted automatically. Only objects performing events are annotated (ID and MBB) in these figures. The good performance of the framework is a result of special considerations to handle inaccuracies and false alarms. For example, the framework is able to differentiate between deposited, split, and obstacle objects.

7. Summary

There are few representation schemes concerning high-level features such as events. Most high-level video representations are context-dependent or focus on the constraints of a narrow application; so they lack generality and flexibility

9. *A computational framework for high-level video representation* 177

Figure 9.17. Key events of the 300 images of the 'Urbicande' shot: Occlusion and (O_1 with O_6, O_1 with O_7, and O_5 with O_7) abnormality (O_1 and O_5).

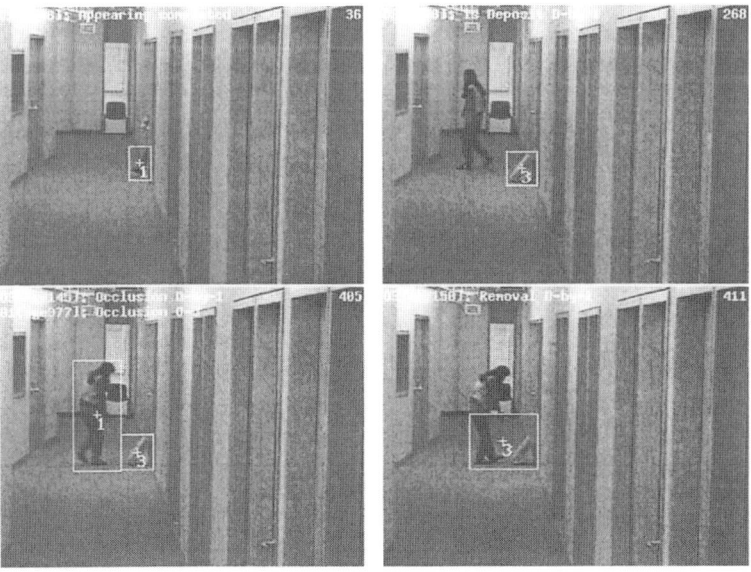

Figure 9.18. Key events of the 'Floor' sequence (826 images): O_1 appears, deposits O_3, moves, and then removes O_3

This paper presented a computational framework to automatically and efficiently extract semantic content from video shots. Semantic content is defined as meaningful video objects and useful context-independent events. Several context independent events have been rigorously defined and automatically detected using features extracted following segmentation, motion estimation and object tracking. The presented events are sufficiently broad to assist applications such as monitoring 1) of removal/deposit of objects, e.g., computing devices, 2) of traffic objects, and 3) behaviors of customers, e.g., in stores. The reliability of the presented framework has been demonstrated by extensive experimentations on indoor and outdoor shots containing over 6000 images with object occlusion, noise, and coding artifacts.

In our framework, special consideration is given to processing inaccuracies and false alarms. For example, the framework is able to differentiate between deposited objects, split objects, and objects at an obstacle. Errors are corrected or compensated at higher level level where more information is available. The presented framework provides a response in real-time for surveillance applications with a rate of up to 10 frames per second.

A critical issue in video surveillance is to differentiate between real moving objects and 'clutter motion', such as trees blowing in the wind and moving shadows. Further research is planned in classification of motion as 'with purpose' (vehicle or people) and 'without purpose' (trees). In addition, the detection of background objects that move during the shot needs to be explicitly processed. Furthermore, a wider set of events will be considered for the framework to serve a larger set of applications.

We are examining further use of our system to 1) overlay tracking information on a display, 2) selectively store video, and 3) tag activities such as cars speeding towards the facility, people climbing walls, and unusual movement around a facility.

Acknowledgments

The author thanks Profs. Eric Dubois and Amar Mitiche for their encouragement and suggestions. This work was supported, in part, by the the Natural Sciences and Engineering Research Council (NSERC) of Canada.

REFERENCES

[1] Aach, T., Kaup, A., and Mester, R. (1993). Statistical model-based change detection in moving video. *Signal Process.*, 31(2):165–180.

[2] Alatan, A., Onural, L., Wollborn, M., Mech, R., Tunceland, E., and Sikora, T. (1998). Image sequence(1) analysis for emerging interactive multimedia services - the European COST 211 Framework. *IEEE Trans. Circuits Syst. Video Technol.*, 8(7):802–813.

[3] Amer, A. (2001). *Object and Event Extraction for Video Processing and Representation in On-Line Video Applications*. PhD thesis, INRS-Télécommunications, Univ. du Québec.

[4] Amer, A. and Dubois, E. (2000). Image segmentation by robust binarization and fast morphological edge detection. In *Proc. IAPR/CIPPRS Int. Conf. on Vision Interface*, pages 357–364, Montréal, Canada.

[5] Amer, A., Mitiche, A., and Dubois, E. (2002). Reliable and fast structure-oriented video noise estimation. In *Proc. IEEE Int. Conf. Image Processing*, volume 1, pages 840–843, Rochester, USA.

[6] Bobick, A.F. (1997). Movement, activity, and action: the role of knowledge in the perception of motion. Technical Report 413, M.I.T. Media Laboratory.

[7] Bouthemy, P., Gelgon, M., and Ganansia, F. (1997). A unified approach to shot change detection and camera motion characterization. Technical Report 1148, Institut National de Recherche en Informatique et en Automatique.

[8] Chang, S.F., Chen, W., Meng, H.J., Sundaram, H., and Zhong, D. (1998). A fully automatic content-based video search engine supporting multi-object spatio-temporal queries. *IEEE Trans. Circuits Syst. Video Techn.*, 8(5):602–615. Special Issue.

[9] Courtney, J. (1997). Automatic video indexing via object motion analysis. *Pattern Recognit.*, 30(4):607–625.

[10] Garrido, L., Salembier, P., and Garcia, D. (1998). Extensive operators in partition lattices for image sequence analysis. *Signal Process.*, 66:157–180.

[11] Haering, N., Qian, R., and Sezan, I. (2000). A semantic event detection approach and its application to detecting hunts in wildlife video. *IEEE Trans. Circuits Syst. Video Techn.*, 10(6):857–868.

[12] Haritaoglu, I., Harwood, D., and Davis, L. S. (2000). W^4: Real-time surveillance of people and their activities. *IEEE Trans. Pattern Anal. Machine Intell.*, 22(8):809–830.

[13] Lippman, A., Vasconcelos, N., and Iyengar, G. (1998). Human interfaces to video. In *Proc. 32nd Asilomar Conf. on Signals, Systems, and Computers*, Asilomar, CA. Invited Paper.

[14] Naphade, M., Mehrotra, R., Ferman, A., Warnick, J., Huang, T., and Tekalp, A. (1998). A high performance algorithm for shot boundary detection using multiple cues. In *Proc. IEEE Int. Conf. Image Processing*, volume 2, pages 884–887, Chicago, IL.

[15] Pentland, A. (2000). Looking at people: Sensing for ubiquitous and wearable computing. *IEEE Trans. Pattern Anal. Machine Intell.*, 22(1):107–119.

[16] Salembier, P., Garrido, L., and Garcia, D. (1997). Image sequence analysis and merging algorithm. In *Proc. Int. Workshop on Very Low Bit-rate Video*, pages 1–8, Linkoping, Sweden. Invited paper.

[17] Smeulders, A.W.M., Worring, M., Santini, S., Gupta, A., and Jain, R. (2000). Content-based image retrieval: the end of the early years. *IEEE Trans. Pattern Anal. Machine Intell.*, 22(12):1349–1380.

[18] Special (2000). Special section on video surveillance. *IEEE Trans. Pattern Anal. Machine Intell.*

[19] Stringa, E. and Regazzoni, C. (1998). Content-based retrieval and real time detection from video sequences acquired by surveillance systems. In *Proc. IEEE Int. Conf. Image Processing*, pages 138–142, Chicago, IL.

[20] Vasconcelos, N. and Lippman, A. (1997). Towards semantically meaningful feature spaces for the characterization of video content. In *Proc. IEEE Int. Conf. Image Processing*, volume 1, pages 25–28, Santa Barbara, CA.

[21] Villegas, P., Marichal, X., and Salcedo, A. (1999). Objective evaluation of segmentation masks in video sequences. In *Proc. Workshop on Image Analysis for Multimedia Interactive Services*, pages 85–88, Berlin, Germany.

[22] Ziliani, F. and Cavallaro, A. (1999). Image analysis for video surveillance based on spatial regularization of a statistical model-based change detection. In *Proc. Int. Conf. on Image Analysis and Processing*, pages 1108–1111, Venice, Italy.

Chapter 10

MULTI-CAMERA SURVEILLANCE
Objec-Based Summarization Approach

Fatih Porikli
Mitsubishi Electric Research Laboratories

Key words: Multi-camera surveillance, summarization, belief-nets

1. INTRODUCTION

Most of the current indoor surveillance applications have single-camera single-room architecture where the cameras are stationary. Typically, each camera is assigned to a dedicated video recorder that can store the streaming video in either time-lapsed or event-based mode. These events are often limited to simple motion detection mechanisms. Considering the huge amount of the video data a multi-camera system may produces over a short time period, more sophisticated tools for control, representation, and content analysis became an urgent need. The nature of surveillance applications demands automatic and accurate detection of object of interest, intra-camera tracking, fusion of multiple modalities to solve inter-camera correspondence problem, easy access and retrieving video data, capability to make semantic query, and abstraction of video content.

Yet another challenge is the extraction of content semantics. In last decade, the coding standards have allowed efficient storage, compression, and communication by handling video information as signals. More recently, object-based encoding and content-based retrieval become possible by extracting and analyzing features of pixels. However, the challenge remains for automatically extracting semantic labels for video content, including labeling of objects, events, places, people, and so forth. By labeling video content at the semantic level, the content will be easier to search, filter, index, summarize, and personalize. The process of video technologies involves transition from dealing with pixels to features to semantics to knowledge as illustrated in Fig. 10-1. In bridging these gaps from pixels to knowledge, objects and models have an important role. The MPEG-7 standard embodies content description models but it does not specify how to extract them. Here, we developed an object-based video content labeling method to restructure the camera-oriented videos into object-oriented results. We propose a summarization technique using the motion activity characteristics of the encoded video segments to provide a solution to the storage and presentation of the immense video data.

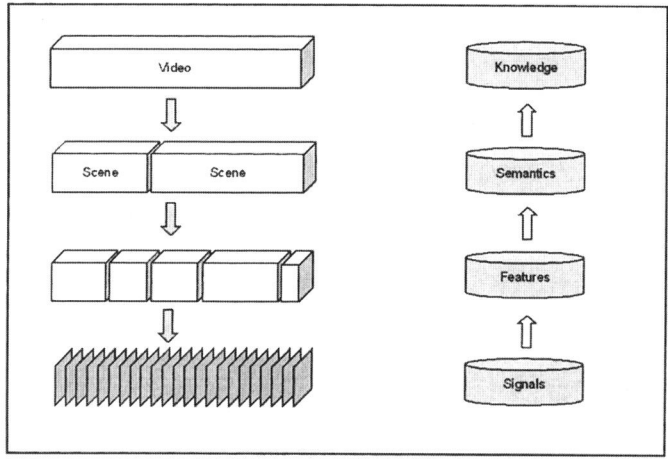

Figure 10-1. Formative and informative representation of a video sequence

Although several multi-camera setups have been adapted for 3D vision problems, the non-overlapping camera systems are not investigated thoroughly. A multi-camera system is proposed in [1]. This system is based on a Gaussian mixture model background subtraction and Kalman filtering to find people in an indoor environment. A Bayesian network is used to combine multiple modalities. Among these modalities, the epipolar, homography, and landmark information assume any pair of cameras in the system has an overlapping field-of-view. Therefore, it is not applicable to the single-camera/single-room architectures.

In this paper, we designed a framework where we can extract the object-wise semantics from a non-overlapping field-of-view multi-camera system. This framework has four main components: automatic tracking, inter-camera data fusion, query generation, and summarization. A flow diagram of the system is shown in Fig. 10-2. In Section 2, we present a single-camera tracking method. Section 3 explains a Bayesian Belief Network for inter-camera correspondence by using object properties and system modalities such as camera location information. In Section 4, we present query generation and summarization.

10. Multi-camera surveillance

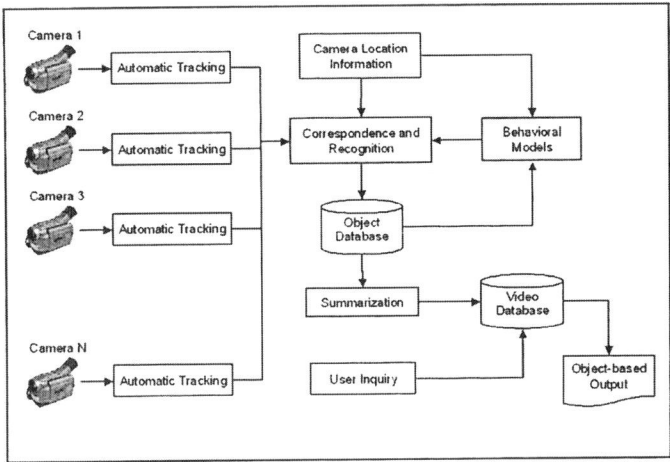

Figure 10-2. Architecture of the multi-camera surveillance system

2. AUTOMATIC TRACKING

A common approach for detecting a moving object for a stationary camera setup is background subtraction. The main idea is to subtract the current image from a reference image that is constructed from the static image pixels during a period of time. Background detection approaches can be classified as non-adaptive and adaptive methods. Manual selection, pixel-wise voting, and mean value search algorithms are among the non-adaptive methods. Adaptive methods include averaging consecutive frames over time, Gaussian mixture models [5,6], alpha blending [7], Kalman filtering [8], and other statistical models [9].

Although averaging and alpha blending are simple and fast, they are not effective for scenes with many moving objects particularly if they move slowly. Besides, they cannot handle multi-modal backgrounds. They may not recover when an object occupies the scene at the initialization phase. Pixel-wise voting among the accumulated images may handle some of the recovery problems, however it becomes computationally very expensive with the increasing number of images.

The Kalman filtering approach may only provide some partial solution. Since lighting conditions may change in most applications, the reference image should be adaptive as well. The Gaussian model based approaches have capability of dealing with illumination changes. Also, it can learn the repetitive variations. However, objects that stop moving may become a part of the background in case the object boundaries are not exact. For the high number of models (>3), this method becomes too slow to be practical.

The tracking of objects can be done either by backtracking or by forward tracking. The backtracking based approach segments foreground regions in the current image and then establishes the correspondence between the current and previous images.

The forward-tracking approach estimates the positions of the regions in the current frame using the segmentation result obtained for the previous image. The limitation of the backtracking approaches is that fixed templates may not be sufficient for all possible objects. A well-known forward-tracking technique is mean-shift analysis, which is a nonparametric density gradient estimator [3]. It is employed to derive the object candidate that is the most similar to a given model while predicting the next object location. This method provides accurate localization, and it is computationally fast. However, it is not automatic since it requires initial models.

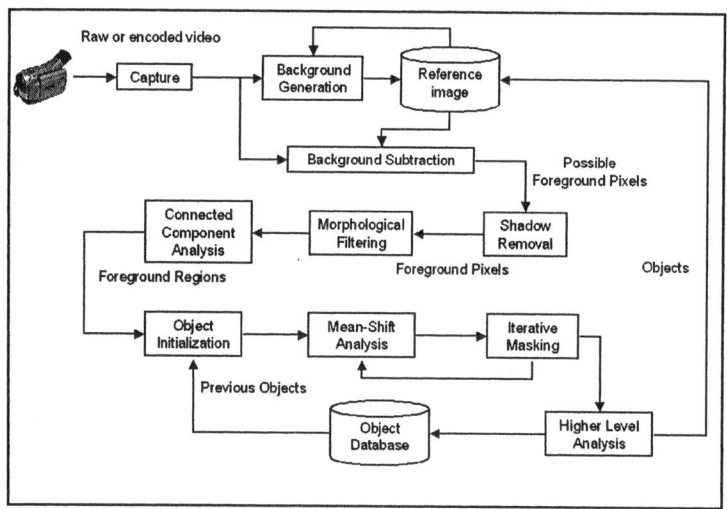

Figure 10-3. Single-camera tracking algorithm.

As shown in Fig. 10-3, our method constructs a reference image using pixel-wise mixture models, finds changed part of image by background subtraction, removes shadows by analyzing color and spatial properties of pixels, determines objects, and tracks them in the consecutive frames.

In background subtraction, the current image is compared to a reference image to detect the changed pixels. The reference image is constructed by utilizing pixel-wise mixture of models as in [5]. We model the history of each pixel by a mixture of Gaussian distributions as

10. Multi-camera surveillance

$$P(p,t) = \sum_{n}^{N} w_n(t) g\left(p, \mu_n(t), \sigma_n^2(t)\right) \tag{10.1}$$

where N is the number of the distributions, $w_n(t)$ is the weight, $\mu_n(t)$ and $\sigma_n^2(t)$ are the mean and the variance of the nth Gaussian model at frame t respectively, and g is a density function as

$$g\left(p, \mu_n(t), \sigma_n^2(t)\right) = \frac{1}{\sqrt{2\pi}\sigma} e^{-\frac{[I(p)-\mu]^2}{\sigma^2}} \tag{10.2}$$

The reference image is updated by comparing the current pixel with the existing Gaussian distributions. In case of the current pixel's color value is within a certain distance of the mean value of a distribution, it is assigned as a match. This distance threshold is set to 2.5σ to include 95% of the distribution. If none of the K distributions ($K<N$) match the current pixel value, a new distribution is initialized. In case of $K=N$, the distribution with the highest variance is replaced with a distribution with the current pixels value as its mean value, and an initial variance. The initial variance is chosen a large value for all distributions. The mean and variance of the matched distributions are updated using a learning coefficient as

$$\begin{aligned}\mu_n(t) &= [1-\alpha]\mu_n(t-1) + \alpha I(p) \\ \sigma_n^2(t) &= [1-\alpha]\sigma_n^2(t-1) + \alpha[\mu_n(t) - I(p)]\end{aligned} \tag{10.3}$$

The weights of the distributions at a frame are adjusted by alpha blending as

$$w_n(t) = \begin{cases} (1-\alpha)w_n(t-1) + \alpha & |\mu_n(t) - I(p)| < 2.5\sigma \\ (1-\alpha)w_n(t-1) & |\mu_n(t) - I(p)| \geq 2.5\sigma \end{cases} \tag{10.4}$$

The learning coefficient α serves as a parameter that controls the rate of the adaptation of the reference image to the current frame. For this purpose, we measure the illumination change δ for a small subset Q of the pixels as

$$\delta = \left| 1 - \sum_{q \in Q} \frac{\langle B(q), I(q) \rangle}{|B(q)|} \right| \tag{10.5}$$

where $B(q)$ represents the background color vector at the pixel q. In case the value of the illumination change is relatively large, the learning parameter is adjusted linearly by $\alpha = c_1 + c_2 \delta$. Unlike the traditional background update mechanisms that refresh the current background model at certain time intervals, we adapt the frequency of the update mechanism Δt_m by using the illumination change as

$$\Delta t_m = \begin{cases} \Delta t_{max} & \delta < \tau_{min} \\ \Delta t_{m-1} & \tau_{min} \leq \delta < \tau_{max} \\ 1 & \tau_{max} \leq \delta \end{cases} \quad (10.6)$$

where $\tau_{min} \ll \tau_{max}$, and Δt_{max} is the number of frames that background model should be updated even if there is no significant illumination change.

After background subtraction, we detect and remove shadow pixels from the set of the foreground pixels. Likelihood of being a shadow pixel is evaluated iteratively by observing the color space attributes and local spatial properties. Shadow removal has two stages. At the first stage pixel-wise color change is evaluated to determine the possible shadow pixels. At the second stage, an iterative classification based on the local information within a local window around a pixel is done. After shadow removal, we have the binary image of foreground pixels that corresponds to the objects. The next task is to find the separate objects. To accomplish this, we first remove speckle noise, then determine connected regions, and group regions into separate objects. To speed up the filtering, we map each 32 horizontal pixels of the binary foreground-background map into a 4-byte integer number. By shifting right and left, and applying logical inclusion with the upper and lower rows, we actually do a morphological dilation operation. In the second pass, logical exclusion is applied similarly to erode the binary image.

While the connected component analysis, we compute the total number of pixels of a connected region, its center of mass, ant its inner/outer boxes coordinates. The inner box contains 90% of the pixels by starting from the pixels close to the center of mass. A rule-based decision mechanism initializes an object by evaluating the connected components. We use box closeness to merge the connected components. For each group of merged components, an object such that its status is set to "possible" is initialized, and a singe outer box is fitted. We track objects by computing the highest gradient direction of color histogram, which is implemented as a maximization process. Using the histogram $h_1(n)$ extracted at the previous frame, this process is iterated as

1. Compute the histogram $h_2(y_0)$ in the current frame, calculate

$$\rho[h_1, h_2(y_0)] = \sum_n^N \sqrt{h_1.h_2(y_0)} \qquad (10.7)$$

2. Compute the weights β_i $i=1,...,R$
3. Derive the new location y_1 by mean-shift,

$$y_1 = \frac{\sum_i^R I(p_i)\beta_i k(p_i)}{\sum_i^R \beta_i k(p_i)} \qquad (10.8)$$

4. Update the target histogram $h_2(y_1)$, and calculate $\rho[h_1, h_2(y_1)]$ as above
5. Stop if $|\rho[h_1, h_2(y_1)]| < \varepsilon$, else $y_1 \rightarrow y_0$, go to step 1.

where the distance between two histograms are defined as $d(y) = \sqrt{1 - \rho[h_1, h_2(y)]}$. A sample tracking result is shown in Fig. 10-4.

Figure 10-4. Single-camera tracking example from MPEG-7 ETRI dataset

3. CORRESPONDENCE AND FUSION

Another issue of the multi-camera system is the problem of integrating the tracking results of multiple cameras. To find the corresponding objects in different cameras and in a central database that keeps records of the previous appearances of all objects, the system evaluates the likelihood of possible pair-wise object matches. This evaluation is done by fusing the general object features such as color, shape, texture features, and other application specific modalities, i.e. camera layout information, behavioral statistics, human-face features, etc.

Color feature is the most common feature that widely accepted by object recognition systems since it is relatively robust towards the size and orientation changes. Possible color features are color templates, histograms, moments, signatures (dominant colors), and partitive color layouts. In a multi-camera setting, illumination, camera distortion, and object resolution differences are most likely to happen. Thus, the color feature should be able to compensate inter-camera distortions as well as illumination changes. Since pixel-wise object template representations are very sensitive to the scale deviations, they are not suitable in our setting. Color signature is defined as a selection of the colors from the quantized color space. A disadvantage of color signature is that they are computational complex. Color layout features have the ability of representing the spatial and color distribution properties at the same time. To extend the global color histogram to a local one, a natural approach is to divide the whole object into sub-blocks and extract color features from each of the sub-blocks. However, they require careful application since they depend on the shape of the object. Thus, we preferred to use color histogram to represent color properties of objects.

Statistically, a color histogram denotes the joint probability of the intensities of the color channels. By modeling the inter-camera distortion and illumination changes as functions of histograms, the matching performance can be improved. We use a cross-correlation based histogram similarity metric to compensate the illumination and inter-camera distortions. This metric uses a cross-correlation matrix H where $h_{mn} = 1 - |h_1(m) - h_2(n)|$ using the normalized color histograms of the corresponding objects at different cameras. We find the maximum gain path by dynamic programming. By comparing this path with an inter-camera characteristic path for the current camera pair, we compensate for the camera distortion. The inter-camera characteristics are obtained by training. This metric evaluates the illumination differences as well.

Texture refers to the visual patterns that have properties of homogeneity that do not result from the presence of only a single color or intensity. Although, it contains important information about the structural arrangement of surfaces and their relationship to the surrounding environment, such level of detail is usually not available in low-resolution surveillance video.

Shape provides another clue for object matching. In general, Fourier descriptor and moment invariants are the most common shape representations. The main idea of Fourier descriptor is to use the Fourier transformed boundary as the shape feature. The main idea of moment invariants is to use region-based moments, which are invariant to transformations, as the shape feature. The biggest drawback of shape features is the sensitivity to scale changes and boundary inaccuracies. Using height descriptor will only help if we have the ground plane. However, since the existing multi-camera systems are difficult to calibrate, a precise ground plane is difficult to obtain. Therefore, the height is not an effective feature.

Using faces to match object between cameras remains the only solution for certain cases, i.e. at a military complex that everyone dresses in identical clothes. However, in typical surveillance applications cameras are usually located far away from the object routes, which result in low-resolution face images. Another concern is the face orientation. Most face-based methods work only for frontal images. The accuracy of identification quickly decreases even with the slight orientation differences. Acquiring a high resolution and frontal picture of an object is not possible always to facilitate facial identification methods. Still, image resolution enhancement techniques may render face features for matching problems in the future.

There is a strong correlation between camera system geometry and likelihood of the objects appearing in a certain camera after they exit from another one. As illustrated in Fig. 10-5, we formulate the camera system as a Bayesian Belief Network, which is a graphical representation of a joint probability distribution over a set of random variables. A BBN is a directed graph in which each set of random variable is represented by a node, and directed edges between nodes represent conditional dependencies. The dependencies can represent the casual inferences among variables. The transition probabilities, which correspond to the likelihood of a person moving from one camera to another linked camera, are learned by observing the system. Note that, each direction on a link may have different probability, however the total incoming and outgoing probability values are equal to one. To satisfy the second constrain, some slack nodes that correspond to the unmonitored entrance/exit regions are added to the graph.

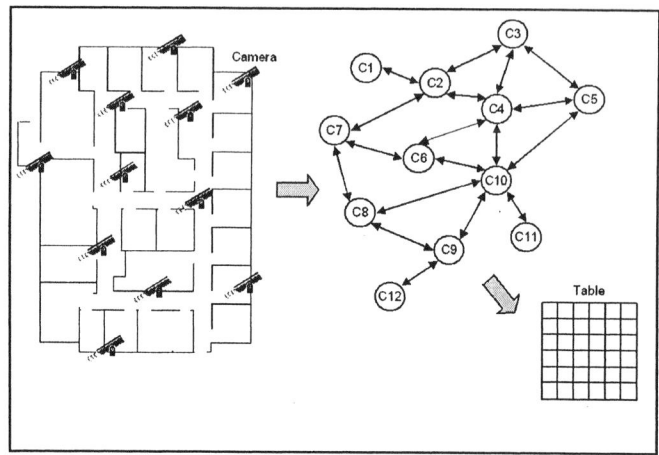

Figure 10-5. Each camera corresponds to a node in the directed graph. The links show the possible physical routes between the cameras.

Initially, there is no object assigned to any node of the BNN, the number of objects in the cameras and objects in the database are equal to zero. The database keeps track of individual objects. Let Ci the camera object O is detected. For each detected new object, a database entry is made using its color histogram features. If the object O exits from the camera Cj, then the conditional probability $P_O(Cj|Ci)\$$ of the same object will be seen on a camera C_j is computed by $P_O(Cj|Ci) = P(Cj|C_{s1})P(C_{s1}|C_{s2})...P(C_{sk}|Ci)$ where $\{s1,s2,..,sk\}$ is the highest probability path from Ci to Cj on the graph. Due to the dynamic nature of the surveillance system, these conditional probabilities should change with time; $P_O(Cj,t|Ci) = P(Cj,t|C_{s1})P(C_{s1},t|C_{s2}...P(C_{sk},t|Ci)$. The conditional probabilities are eroded by time as $P_O(Cj,t|Ci) = k.P_O(Cj,t-1|Ci)$ where $k<1$ since the object may exit from the system completely. Here, we do not think a multi-camera system should be a closed graph. However, the conditional probabilities do not become less than a threshold $P_O(Cj,\infty|C_i) = 1/(M+1)$, which corresponds to the identical and independent nodes. Here, M is the number of cameras, and the addition is due to we treat the database as another node.

As a new object is detected, it is compared with the objects in the database and with the objects disappeared from a camera but still is not matched. The comparison is based on the color histogram similarity. For more than one object correspondence, we normalize each similarity score with the total of similarity scores of all possible pairs. By scaling these scores with the conditional probabilities, we select the correct match as the pair (Om,On) that maximizes $g^*_{mn} P(Cm,t|Ci) P(Cn,t|Cj)$. To match objects between two cameras Ci and Cj, we evaluate the matching for all objects simultaneously instead of matching each single subset independently to minimize the matching conflicts.

4. OBJECT-BASED QUERY AND SUMMARIZATION

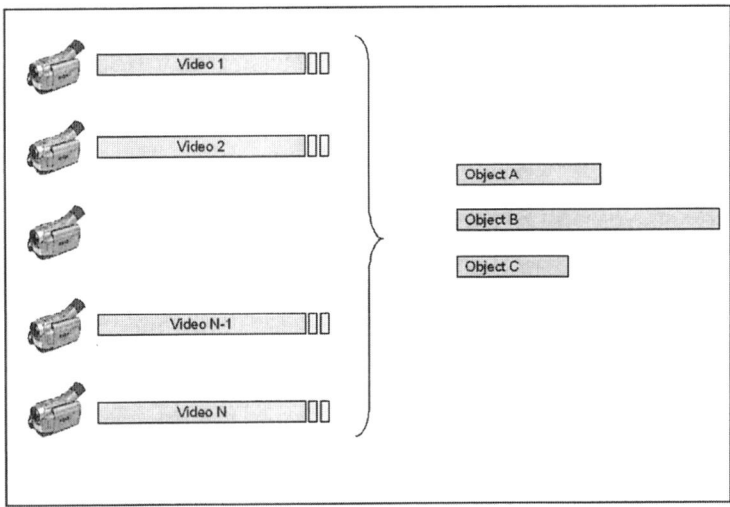

Figure 10-6. Instead of camera-based representation, multiple videos can be restructured as object-based sequences

After matching the objects between the cameras, we label each video frame according to the object appearances. This enables us to include content semantics in the subsequent processes. A semantic scene is defined as a collection of shots that are consistent with respect to a certain semantic, in our case, the identities of objects. Since we have the position information of the cameras and we extracted which objects appeared in a which video at what time, we can, for instance, query for what an object did in a certain time period at a certain location. Thus, we are able to represent the video sequences with respect to the detected objects as illustrated in Fig. 10-6.

To obtain concise and very low-bit representation of query results, we generate an abstract of the query result video sequence. Video abstraction can be done either by providing a ``preview'', which consists of a concatenation of key-video segments, or a set of key frames chosen from the frames that comprise the video sequence. The key-frame-based summary is a collection of frames that aims to capture all of the visual essence of the video except, of course, the motion. Making it ideal for rapid browsing of stored video, its constituent key frames can serve as pointers to the desired portions of the content.

A key frame generation technique based on a measure of the fidelity of a set of key frames is introduced in [2]. The fidelity measure is defined as the Semi-Hausdorff distance between the set of key frames S and the set of frames R in the video sequences. A practical definition of the Semi-Hausdorff distance is as follows: Let the key frame set consist of S_{max} frames, and let the set of frames R contain R_{max} frames. Let the dissimilarity between two frames Si and Ri be $d(Si,Ri)$. We define *fi* for a frame Ri as *fi=min[d(Sk,Ri)]*, $k=1..S_{max}$. Then the Semi-Hausdorff distance between S and R is given by *max[fi]*, $i=1..R_{max}$. This way we end up finding out how well the key frame set S represents R, because the better the representation the lower the Semi-Hausdorff distance between S and R. For example, in the trivial case, if the S and R are identical, the Semi-Hausdorff distance is zero. On the other hand, a high Semi-Hausdorff distance indicates that at least one of the frames in R was not well represented by any of the frames in the key frame set S. As a frame dissimilarity metric, we proposed to use motion activity descriptor. The motion activity score of a frame is defined as the standard deviation of motion vector magnitude. By treating the motion activity scores of a video segment as a distribution function, we obtain a cumulative motion activity function. We quantize standard deviation of motion vectors of MPEG-1 video to classify segments into five classes ranging from very low to very high intensity. In [4], it is showed that the frame at which the cumulative motion activity is half the maximum value is also the halfway point for the cumulative increment in information. This implies that it would be the best choice for the first key frame since it would have the minimum Semi-Hausdorff distance. Being forced to pick the first frame as a key frame is disadvantageous, i.e. not all of the target object is visible, or it is very small in the first frame in comparison to rest of the segment. This motivates us to find a key frame based on motion activity that would be better than the first frame. Thus, we select the key frames according to the cumulative function as shown in Fig. 10-7.

10. Multi-camera surveillance

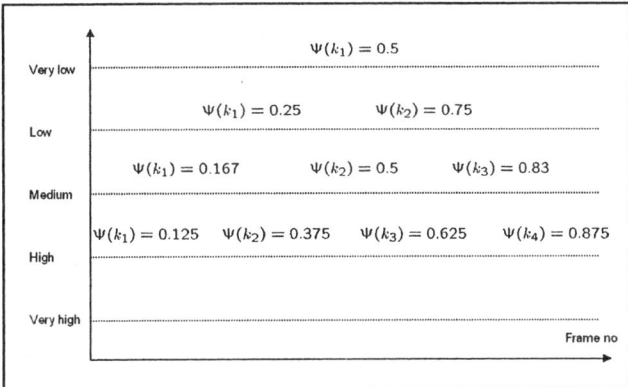

Figure 10-7. Motion activity based key-frame extraction optimal strategy: Note that the middle of each segment is picked

For a single camera, tracking takes less than 28 milliseconds/frame on average for color video at 320x240 spatial resolution on a Pentium4, 1.8Ghz computer. A base station controls the fusion of the multi camera information, and its computational load is negligible. We presented the object-based inquiry user interface in Fig. 10-8. The tracking method can follow several objects at the same time even some of the objects are totally occluded for a long time. Furthermore, it provides accurate object boundaries. We initialized the Bayesian Network with identical conditional probabilities. These probabilities may also be adapted by observing the long-term motion behavior of the objects. Sample query results are shown in Fig. 10-9. We are able to extract all appearances of the same person in a specified time period accurately. We can also count the number of different people. The background generation method is computationally more feasible than the existing mixture models, and it can achieve real-time performance even for full resolution video owing to the new illumination change detection and reference image refresh mechanisms. The shadow removal method effectively filters most shadow pixels without breaking object regions apart. This method is robust towards the perturbations of the filter parameters, and it adapts easily for different lighting conditions. The performance of the background adaptation and mean-shift analysis based object tracking method is comparable with the state-of-art, and it is fully automatic. It does not have the intrinsic shortcomings of the template-matching approaches such as resolution, pose, and illumination dependencies. The object-based representation enables us to associate content semantics, thus we can generate query-based summaries. This is an important functionality to retrieve the target video segments from a large database of surveillance video. The motion activity based summarization is numerically and visually comparable with the existing techniques and relies on computationally simple motion feature extraction in the compressed domain, and is thus much simpler than other techniques. Using additional biometrics such as face, gait, and speech, may be necessary for long-time tracking scenarios where the color features of an object change. Face recognition is a possible solution for this problem. Currently, we are investigating robust and computationally feasible ways of integrating face features.

One storage space related challenge of the object-based representation arises when the object number is much greater than the number of cameras. In this case, instead of storing object-based video sequences, a conversion table that keeps the pointers from the camera-based video segments to the object-based video segments may be a better solution. The current tracking system is designed for stationary cameras. In the future, we consider improving the object tracking method so that it can handle pan-tilt-zoom cameras as well.

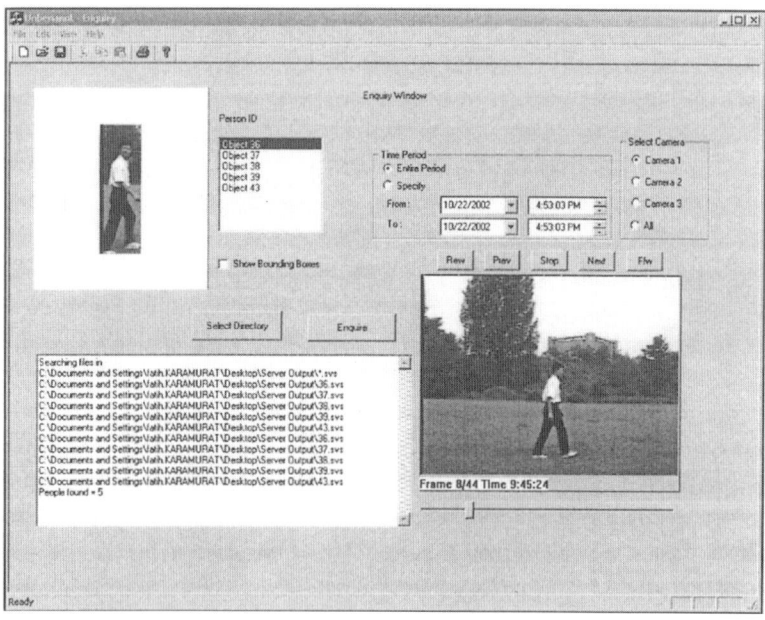

Figure 10-8. Inquiry system that a user can specify the camera, time, and object to generate summary

10. Multi-camera surveillance

Figure 10-9. The retrieved instances of two objects in one camera; the person with white shirt at frames 10, 20, 24 (first row), 215, 224, 238 (second row), and another person with red shirt 142, 154, 392 (last row) on the same camera.

REFERENCES

[1] T. H. Chang and S. Gong. ``Bayesian modality fusion for tracking multiple people with a multi-camera system". *In Proc. European Workshop on Advanced Video-based Surveillance Systems*, Kingston, UK, 2001

[2] H. S. Chang, S. Sull, and S. U. Lee, "Efficient video indexing scheme for content-based retrieval,'' *IEEE Trans. Circuits Syst. Video Technol.*, Vol. 9, no. 8, 1999.

[3] D. Comaniciu, V. Ramesh, and P. Meer, "Real-time tracking of non-rigid objects sing mean shift, *IEEE Conf. Computer Vision and Pattern Recognition*, SouthCarolina, Vol. 2, 142-149, 2000.

[4] A. Divakaran ``Video summarization using descriptors of motion activity: A motion activity based approach to key-frame extraction from video shots", *Journal of Electronic Imaging*, Vol. 10, no. 4, 2001.

[5] Friedman and S. Russell, "Image segmentation in video sequences: A probabilistic approach," In *Proc. of the Thirteenth Conf. on Uncertainty in Artificial Intelligence*, 1997.

[6] C. Stauffer and W.Grimson, "Adaptive background mixture models for real-time tracking", *Proc. IEEE Conf. on Computer Vision and Pattern Recognition*, Vol. 2, 1999.

7. T.Horprasert, D. Harwood, and L. Davis, "A statistical approach for real-time robust background subtraction and shadow detection", *Proc. of IEEE Intern. Conference on Computer Vision, FRAME-RATE Workshop*, 1999.

8. C. Ridder, O. Munkelt, and H. Kirchner, ``Adaptive background estimation and foreground detection using Kalman-filtering", *Proc. of Int. Conf. on recent Advances in Mechatronics*, 193-199, 1995.

9. C.R. Wren, A. Azarbayejani, T. Darrell, and A. Pentland, "Pfinder: Real-time tracking of the human body", *IEEE Transactions on Pattern Analysis and Machine Intelligence*, Vol. 19, no. 7, pp.780–785, July 1998.

Chapter 11

DETECTING DANGEROUS BEHAVIORS OF MOBILE OBJECTS IN PARKING AREAS

Gian Luca Foresti, Giorgio Giacinto[*] and Fabio Roli[*]
Department of Mathematics and Computer Science, University of Udine, Via delle Scienze 206, 33100 Udine, ITALY

[*]*Department of Electronic Engineering, University of Cagliari, Piazza d' Armi, 09123 Cagliari, ITALY*

Key words: Visual surveillance, parking lots, neural networks, learning algorithm, trajectory analysis, object tracking

1. INTRODUCTION

In the last decade video-surveillance systems have been developed for monitoring remote environments in order to detect and prevent dangerous situations. Until few years ago, surveillance was performed entirely by human operators, who interpreted the visual information presented to them on one or more monitors.

Sometimes, their work conditions resulted in failing to raise the generation of alarms in case of dangerous situations. More recently an effort has been made to develop automatic or semi-automatic video-based surveillance systems [1], able to alert human operators when something unusual occurs within the monitored environment. Such systems were developed for different tasks, such as object detection [2,3], vehicle tracking on highways for traffic control purposes [4-6], analysis of human behaviour [7-10] or people counting [11] in public environments. The development of complex surveillance systems is capturing interest of both research and industrial world as there are strong requirements coming from the society in the direction of increasing safety and security in different applications.

An automatic surveillance system can be defined as a computer system that, based on input data and provided by scene understanding capabilities, is oriented to either human operators attention focusing or to automatic alarm generation. Video-surveillance systems have been developed since 60s:

First generation (1960-1980) - in these kind of systems there is no information processing; acquired images are presented on one or more monitors, and the scene understanding task is performed by the human operator.

Second generation (1990-2000) - digital information processing is completely performed by a centralized system that shows acquired scenes and generates alarms helping the human operator to face potentially dangerous situations.

Third generation (2000-) - distributed digital signal processing. Only the information for signalling and describing a dangerous situation is presented to the human operator.

Sensors acquiring input data from the outside world to the system can be heterogeneous (TV cameras, IR cameras, microphones, GPS sensors, , etc.) and arranged in multisensor configurations. Multisensor data are processed in order to obtain a complete understanding of the observed scene. Scene understanding requirements regarding a video-based surveillance system are severe due to high variability and poor structure of monitored scenes in different applications. Such variability has several consequences in required processing tools. For example, the use of sophisticated image processing algorithms is necessary for rough signal pre-processing and filtering. Change detection, based on the difference between two video frames, is one of the most efficient techniques for detecting moving objects [12]. However, some problems arise when an object moving in the scene stops in a given point. Some researchers have suggest the use of dynamic background frame differencing to overcome this difficulty [3]. Several sophisticated approaches have been designed for object tracking. They required combinations of 2-D and 3-D models [13], mesh-based constraints [14], deformable contours [15], motion estimation [16]. Davies et al. use a combination of wavelets filtering and a Kalman-based algorithm for tracking a number of small, low contrast object through image sequences taken with a static camera [17]. Smith introduces a real-time system for motion segmentation and object tracking in complex scenes using a variation of deformable contour tracking [18].

Highly variable scene conditions imply the necessity of selecting robust scene description and pattern recognition methods. Automatic learning capabilities are an emerging issue in surveillance systems, as the capability of automatically developing models of scenes to be recognized as potentially dangerous events from a training set of presented examples will be a key issue for improving end-user acceptance of surveillance systems. Recently, some researchers have provided some solutions to the problem of event detection and scene understanding. Buxton et al. introduced Bayesian networks to detect interesting events into a dynamic scene and to provide interpretation of traffic situations [19]. More recently, Bobick proposed a new approach for the description and interpretation of human activities in the context of a visual surveillance task [20]. Foresti and Roli [21] present a visual-based surveillance system for real-time event detection and classification. The system is able to interpret some events, generated by human moving in a road scene, by classifying them as normal or potentially dangerous. In particular, a neural tree network is learned with statistical models of human trajectories pre-classified as typical or atypical.

In this paper, the work proposed in [21] is extended to more complex outdoor scenes, i.e., parking areas, where an human or a vehicle can assume several and different behaviors. In this context, the number of possible events (normal and/or dangerous) that occur in the observed scene is very high, so the learning process of the system from real situations can become very time consuming and tedious. In order to face this problem, the use of recognition techniques based on a "learning by example" paradigm has been considered. To this end, a statistical model of the distribution of "normal" and "dangerous" trajectories based on a relatively small set of examples, i.e., a set of "prototypal" trajectories manually selected, has been obtained. Then, the event recognition problem has been formulated in terms of a "pattern recognition" problem, where the trajectories, represented by a Beziér curve, play the role of "patterns" that must be assigned to one of the two data classes, e.g. "normal" or "dangerous" trajectories. Patterns are extracted from the image sequence by estimating a set of local directions of the moving object at some points. Finally, the whole set of sample trajectories is stored into an event database and used to train a multi-layer perceptron (MLP) neural network.

2. SYSTEM DESCRIPTION

Figure 11-1 shows the general architecture of the proposed surveillance system.

A stationary CCD camera is used to acquire image sequences of the monitored scene. A change detection module (CD) [3] is applied to each frame $I(x,y)$ of the input image sequence and it makes out a binary image $B(x,y)$ where each blob represents a possible object (e.g., vehicle, van, trunk, motorcycle, people) moving in the scene. A background updating procedure based on Kalman filters is applied to estimate significant changes of the background scene [12]. A localization and tracking module, which uses as input geometric information about the blobs, is applied to estimate the motion parameters of the detected objects [16]. A camera calibration procedure [14] and the ground-plane hypothesis [5] are considered. A neural tree network [21,22] is applied to classify each detected blob among a predefined set of object models (e.g., pedestrians and cars have been selected).

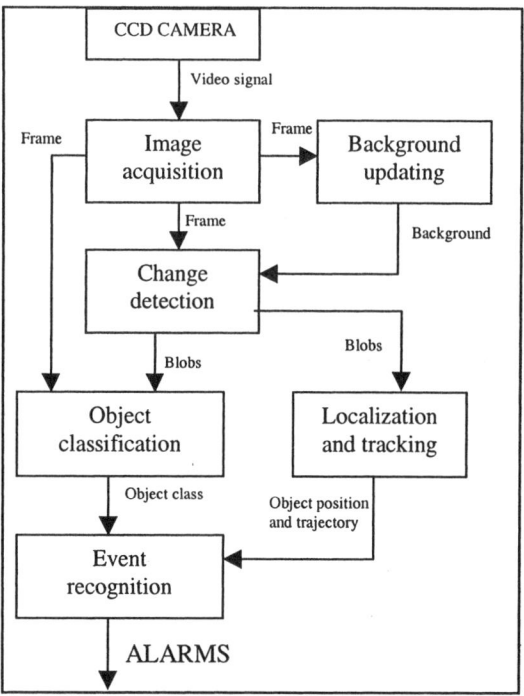

Figure 11-1 – General architecture of the surveillance system.

Finally, the event recognition module integrates information over N consecutive frames about the class and the trajectory of the detected object to classify its behaviour as normal or dangerous. This module is represented by a a multi-layer perceptron (MLP) neural network trained by using the error back-propagation (EBP) learning algorithm.

3. TRAJECTORY RECOGNITION

The recognition task at hand is related to the parking area shown in Fig. 11-2. Two types of moving objects are considered: people and cars. We assume that the presence of moving objects other than cars and people is automatically filtered by the pre-processing stage. Thus we concentrate on the interpretation of the behaviour of either pedestrian or cars, under the assumption that their trajectories are acquired according to the techniques outlined in the previous section. In particular, our task focused on the recognition of potentially dangerous events such as (*i*) vehicles or pedestrian moving along the road with a non rectilinear trajectory, (*ii*) vehicles which cross the lane, (*iii*) pedestrians walking on the road outside the zebra crossing, etc.

(a)

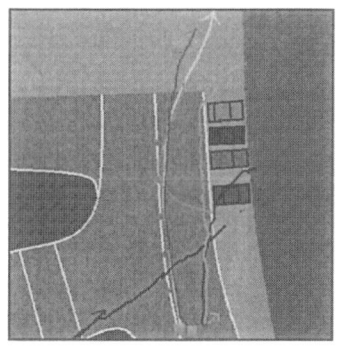

(b)   *(c)*

Figure 11-2 - The parking area chosen as a case study: (a) original image taken by the camera; (b) the plant used for the analysis of trajectories, where two zebra crossing have been assumed; (c) some real trajectories of cars (characterized by a small coloured rectangle at the beginning of the line) and pedestrians.

Figure 11-3a shows some examples of typical trajectories of cars and pedestrians (we have assumed the existence of two zebra crossing, showed in Fig. 11-2b. However they have been reported in other figures for the sake of simplicity), while Figs. 11-3b and 11-3c show some examples of dangerous trajectories of pedestrian and cars respectively. It is worth noting that the number of possible trajectories, either typical or dangerous ones, is very large even for a simple scene like the considered case study.

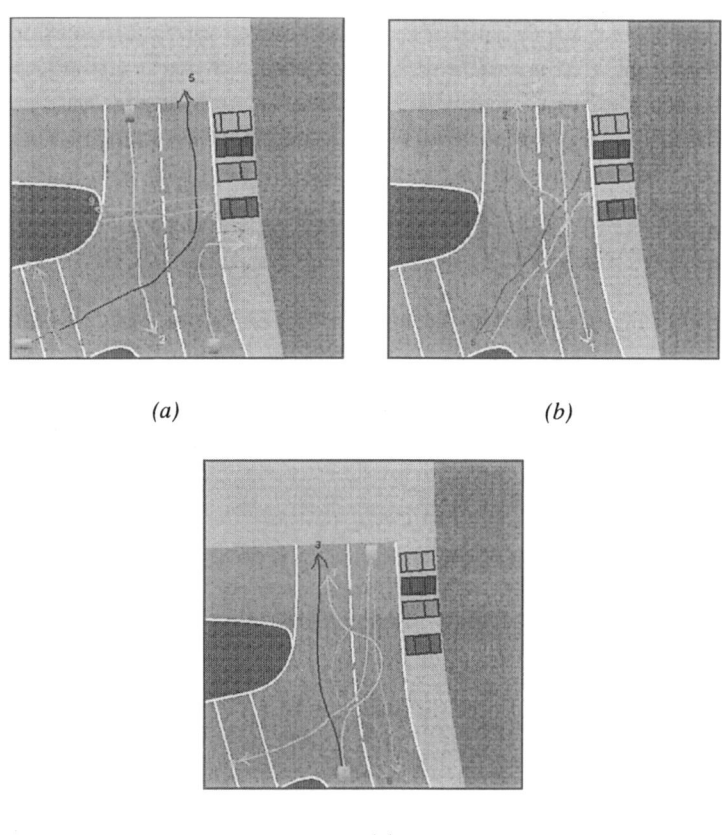

Figure 11-3 - Examples of trajectories of pedestrians and cars in our case study: (a) normal trajectories; (b) dangerous trajectories of pedestrians; (c) dangerous trajectories of cars

In order to design a recognition tool that can distinguish between dangerous trajectories and normal ones, we think that two alternatives can be followed [23]:

– The first method involves the use of hand generated models of object behaviours, under the assumption that objects tend to move in predefined ways;

- The second method relies on a learnt model of typical and atypical object behaviours. Such a model should exhibit some generalization capability in order to recognize dangerous events in cases where all possible object behaviours cannot be predefined.

The case study at hand is an example of the second situation since it is not possible to predefine all possible trajectories of pedestrians and cars. On the other hand, we are able to design a representative set of typical and atypical trajectories. Consequently, our approach involves the use of recognition techniques based on a "learning by example" paradigm. The result of the learning phase is a statistical model of the distribution of "normal" and "dangerous" trajectories based on a relatively small set of examples.

We formulate the event recognition problem in terms of a "pattern recognition" problem, where the trajectories play the role of "patterns" that must be assigned to one of the two data classes, e.g. "normal" or "dangerous" trajectories. Under the pattern recognition framework each pattern must be described by a set of suitable "features" or attributes. Usually features are selected among the set of measures that can be used to describe an object. In particular, features are those measures which can be assumed to have some discriminative power for the recognition task at hand [24]. The task of classifying a trajectory as "normal" or "dangerous" can be performed by describing each trajectory by a set of local directions [21]. It is easy to see that for a given scene "normal" and "dangerous" trajectories can be identified by estimating a set of local directions of the moving object at some points. In many cases this can be considered a suitable approximation of an object trajectory. In particular, the trajectory analysis module uses the Beziér curves to approximate the object trajectory by starting from a set of of N consecutive object positions. Let $\mathbf{B}_0,\ldots,\mathbf{B}_N$ be (N+1) points belonging to the road (ground) plane. The parametric Beziér curve generated by the above (N+1) points is given by

$$\mathbf{P}(t) = \sum_{i=0}^{n} \mathbf{B}_i J_{n,i}(t) \qquad 0 \leq t \leq 1 \tag{11.1}$$

where

$$J_{n,i}(t) = \binom{n}{i} t^i (1-t)^{n-i} \tag{11.2}$$

The following steps are necessary to compute the local direction at a given point:

(a) computation of the first derivative of the curve $P(t)$

$$\mathbf{P}'(t) = [P'_x(t), P'_y(t)] = \sum_{i=0}^{n} \mathbf{B}_i J'_{n,i}(t) \qquad (11.3)$$

where

$$J'_{n,i}(t) = \frac{(i-nt)}{t(1-t)} J_{n,i}(t) \qquad (11.4)$$

(b) determination of the object direction in a given point

$$d(t) = \tan^{-1}\left(\frac{P'_y(t)}{P'_x(t)}\right) \qquad (11.5)$$

Let us consider a tracked object, i.e. a car or a pedestrian, at frame n. Let us denote with $Q(n)$ the feature vector associated with this object. The feature vector $Q(n)$ can be filled up by a set of points of the Beziér curve $P(t)$ obtained by sampling $P(t)$ at different t values. For each point three features are stored in $Q(n)$: its $(x(t),y(t))$ position in the plane and the direction $d(t)$ of the trajectory. For example, the set of points can be extracted from the trajectory by sampling the related Beziér curve at regular intervals.

The whole set of sample trajectories is stored into an event database and used to train a multi-layer perceptron (MLP) neural network. Neural classifiers are currently adopted in advanced video-surveillance systems to classify objects in detected blobs thanks to their ability to solve complex classification tasks [22]. Thus neural networks have been also used to perform the event recognition task. MLP neural networks training has been carried out using the error back-propagation (EBP) learning algorithm.

4. EXPERIMENTAL RESULTS

In order to test the system presented in this paper, two main activities have been performed. The first activity was the definition of a suitable set of examples of pedestrian and car trajectories for the parking area at hand. This set of examples has been subsequently subdivided into two subsets, one of them used for training the MLP neural network and the other one used for the evaluation of the system performances.

4.1 Data set generation

It is worth noting that few training data related to dangerous situations can be extracted from real scenes since it is easy to see that usually object trajectories are normal. Moreover it could take several weeks of recording to collect a set of trajectories that can be reasonably considered as representative of all the possible object trajectories. In order to overcome this difficulty, we propose to create manually a set of representative trajectories. In particular we considered a small set of 20 "prototypal" trajectories (see Fig. 11-2) that covers a wide range of different object behaviours. 10 prototypes are related to a "normal" behaviour, while the other 10 are related to "dangerous" behaviours (5 of them related to cars and the others to pedestrians). It is worth noting that our purpose was not to create all possible trajectories, but only a representative subset.

Each "prototypal" trajectory has been then used as input to a procedure designed to automatically generate a number of similar different trajectories. For each prototypal trajectory, the points used to generate the Bezier curve has been considered. New trajectories have been then generated by moving such points in the (x,y) plane. These movements has been constrained so that a prototypal "typical" trajectory generates new "typical" trajectories, while prototypal "dangerous" trajectories generate only "dangerous" trajectories. Moreover the constraints allowed a new trajectory to be obtained from the prototype by rotation and translation. Figure 11-4 shows an example of this procedure.

For each prototype we generated 15 new trajectories and the final data set contained 300 trajectories. This data set has been randomly subdivided into two sets, keeping the proportion between normal and dangerous trajectories equal to the one of the whole set. One of the two sets has been used for MLP network training and the other one for testing.

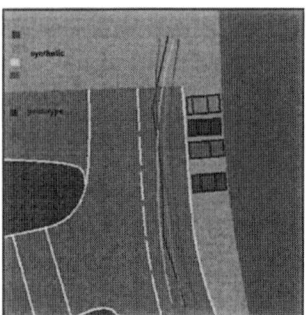

Figure 11-4 – A prototype trajectory and some of the synthetic trajectories that have been generated

4.2 Event recognition results

The above data set has been used to train a MLP neural network. Since the goal of the classification procedure is to recognize an event as normal or dangerous, the problem can be formulated as a two class problem. Thus it follows that an MLP with one output neuron can be used. Each input trajectory has been sampled at 9 points corresponding to values of the t parameter from 0.01 to 0.99 with a constant step equal to 0.1225. According to Section III, three features has been associated with each point, namely the values of x, y, and $d(t)$. Consequently 27 features are available for each trajectory. An additional feature has been considered to distinguish between pedestrian trajectories and car trajectories. Therefore the architecture of the MLP had 28 input feature and one output neuron. In our experiments we considered an hidden layer made up of 28 neurons.

Several experiments has been carried out using different learning parameters in order to assess the best performances. In addition different training sets containing different proportions of "normal" and "dangerous" trajectories have been considered in order to simulate a real scenario where the number of normal events is higher than the number of "dangerous" events. Results are reported in Table 11-1 in terms of percentage of false alarms and missed alarms on the test set. The percentage of false alarms is always negligible, thanks to the fact that the training set always contains enough examples of normal trajectories. Therefore normal trajectories can be hardly classified as alarms. On the contrary the percentage of missed alarms grows as a smaller percentage of dangerous events is included in the training set. These performances are in agreement with any "learning by example" paradigm where a poor represented class of examples results in poor performances on the test set.

Table 11-1 – False and missed alarms obtained by the system on the test set. Different training set containing different proportions of normal and dangerous trajectories have been considered.

	Percentage of Dangerous Trajectories in the Training Set					
	5%	10%	20%	30%	40%	50%
% False Alarms	0	0	2	1.85	1.39	1.32
% Missed Alarms	74.67	36	34.67	29.33	5.33	0

The overall performances of the systems for training sets containing more than 30% of dangerous events are quite good. It is worth recalling that the recognition task at hand is quite challenging, since normal trajectories and dangerous ones are often very similar each other (see for example the trajectories showed in Fig. 11-3). Thus it can be concluded that the proposed approach provides an effective solution to the event recognition problem.

5. CONCLUSIONS

A surveillance system for detecting dangerous events in a parking area scene has been presented. The ability of the system is not only restricted to detect, track and recognize mobile objects, but also to detect and recognize some dangerous events, e.g., objects moving in the scene with dangerous behaviour.

The system uses a neural tree to perform object classification and an MLP neural network to perform event recognition. In order to overcome the lack of enough data for neural network training and testing, a synthetic data set has been generated from a small set of real data. The system provided a small percentage of false alarms and missed alarms thus suggesting that the proposed methodology is an effective solution to the event recognition problem at hand. Further results will be obtained by testing the proposed system on a larger training set containing both real and synthetic data.

ACKNOWLEDGMENTS

This work was partially supported by the Italian Ministry of University and Scientific Research within the framework of the project *"Distributed systems for multisensor recongnition with augmented perception for ambient security and customization"* (2002-2004).

REFERENCES

[1] G.L. Foresti, P. Mahonen and C.S. Regazzoni (eds.) *Multimedia Video-Based Surveillance Systems: Requirements, Issues and Solutions*, Kluwer Academic Publishers, Boston, 2000.

[2] S. Ulman, *High-level vision–Object recognition and visual cognition*, MIT Press, 1996.

[3] G.L. Foresti, "Object detection and tracking in time-varying and badly illuminated outdoor environments", *Optical Engineering*, Vol. 37, No. 9, 1998, pp. 2550-2564.

[4] A.F.Toal and H.Buxton, "Spatio-temporal reasoning within a traffic surveillance system," in *Proceedings of 2th European Conference on Computer Vision*, S. Margherita, Italy, 1992, pp. 884-892.

[5] D. Koller, K. Daniilidis, and H. Nagel, "Model-based object tracking in monocular image sequences of road traffic scenes," *International Journal of Computer Vision*, vol. 10, pp. 257-281, 1993.

[6] Blosseville, J.M. (1999), "Image Processing for Traffic Management", in C.S.Regazzoni, G. Vernazza, and G. Fabri (eds.), *Advanced Video-Based Surveillance System*s, Kluwer Academic Publishers, Norwell, pp. 67-75.

[7] J. K. Aggarwal and Q. Cai, "Human Motion Analysis: A Review", *Computer Vision and Image Understanding*, vol. 73, no. 3, 1999, pp. 428-440.

[8] K. Rohr, "Towards model-based recognition of human movements in image sequences", *Computer Vision, Graphics and Image Processing: Image Understanding*, vol. 59, no. 1, 1994, pp. 94-115.

[9] L.Q. Xu and D.C. Hogg, "Neural networks in human motion tracking – an experimental study", *Image and Vision Computing*, vol. 15, no. 8, 1997, pp. 607-615.

[10] R. Bowden, T.A. Mitchell, and M. Sarhadi, "Non-linear statistical models for the 3D reconstruction of human pose and motion from monocular image sequences", *Image and Vision Computing*, vol. 18, no. 9, 2000, pp. 729-737.

[11] Regazzoni, C.S. and Tesei, A., (1996) Distributed Data Fusion for Real-Time Crowding Estimation, *Signal Processing*, Vol.53, pp.47-63.

[12] T. Aach and A. Kaup, "Bayesian algorithms for adaptive change detection in image sequences using Markov random fields", *Signal Processing*, Vol. 7, 1995, pp. 147-160.

[13] Z. Li and H. Wang, "Real-time 3-D motion tracking with known geometric models", *Real-Time Imaging*, Vol. 5, 1999, pp. 167-187.

[14] P.J.L. Van Beek, A.M. Tekalp, N. Zhuang, I. Celasun, M. Xia, "Hierarchical 2-D mesh representation, tracking and compression for object-based video", *IEEE Transaction on Circuits and Systems for Video Technology*, Vol. 9, 1999, pp. 617-634.

[15] F. Bremond and M. Thonnat, "Tracking multiple nonrigid objects in video sequences", *IEEE Trans. on Circuits and Systems for Video Technology*, Vol. 8, 1998, pp. 585-591.

[16] G.L. Foresti, "Real-time detection of multiple moving objects in complex image sequences", *Int. Journal of Imaging Systems and Technology*, Vol. 10, 1999, pp. 305-317.

[17] D. Davies, P. Palmer and M. Mirmehdi, "Detection and tracking of very small low contrast objects", in *Proceedings of the 9th British Machine Vision Conference*, 1998, pp. 599-608.

[18] S.M. Smith, "ASSET-2: real-time motion segmentation and object tracking", *Real Time Imaging*, vol. 4, 1998, pp. 21-40.

[19] H. Buxton and S. Gong, "Visual surveillance in a dynamic and uncertain world", *Artificial Intelligence*, Vol. 78, No. 1-2, 1995, pp. 431-459.

[20] A.F. Bobick, "Computer seeing action", in *Proceedings of the 7th Annual British Machine Vision Conference,* 1996, pp. 13-22.

[21] G.L. Foresti and F. Roli, "Real-time recognition of suspicious events for advanced visual-based surveillance" in *Multimedia Video-Based Surveillance Systems: Requirements, Issues and Solutions*, G.L. Foresti, P. Mahonen and C.S. Regazzoni and (eds.), Kluwer Academic Publishers, Norwell, 2000, pp. 84-93.

[22] G.L. Foresti, "Outdoor Scene Classification by a Neural Tree Based Approach", *Pattern Analysis and Applications*, Vol. 2, 1999, pp. 129-142.

[23] N. Johnson, D. Hogg, "Learning the distribution of object trajectories for event recognition", *Image and Vision Computing*, Vol. 14, 1996, pp. 609-615.

[24] K. Fukunaga, *Introduction to Statistical Pattern Recognition*, Academic Press, 1990.

III

BIOMETRICS IN SURVEILLANCE SYSTEMS

BIOMETRICS IN SURVEILLANCE SYSTEMS

Carlo S. Regazzoni
Department of Biophysical and Electronical Engineering - University of Genova, ITALY

In this chapter, the use of biometrics in automatic video and signal based systems is explored. Surveillance systems based on biometrics are becoming more and more important at this time. The reason lies in the contextual historical moment where the demand for security tools as related to potentially dangerous single human actions is increasing. Another motivation is related to other emerging functionalities related to the availability of an extended and distributed amount of computational and communication resources, eventually embedded in intelligent sensors. Such capabilities seem to make it possible in a middle term future at accessible costs ubiquitous multimodal interfaces able to recognize humans and to adaptively interact with them, hopefully, but depending also on humans behaviour itself, in a friendly manner. The problem of assessing the identity of an individual and of maintaining it in a coherent framework along time and space can be solved if static and dynamic biometrics can be established.

In this chapter, four papers are presented that make it understandable the amount of research effort towards recognizing humans and their behaviours in the most precise and complete way possible. The first three papers deal with feature extraction and classification of video images and sequences where humans are observed. They deal explicitly with the capability of tracking and recognizing human bodies and faces within a statistical measurement and decision tool. Even though this topic has been extensively considered in literature, it remains still an open problem, especially due to the wide variability of the external environment conditions (including observed humans) as perceived under different sensor set-up. As a consequence, robustness and adaptability necessary for example for a 24-hours outdoor system still have to be reached at a sufficient performance level. On the other side, the fourth paper provides a system viewpoint by addressing the problem that will be a key for success for security systems emerging on the market: data fusion intended here as the capability of designing optimal scalable rules for systems using heterogeneous sensors to monitor the same event such as person identity. Fusion of multiple sensors observations (e.g. voice, face, and fingerprinting recognisers) for assessing the identity of an individual is considered as a sample problem, but presented results can be surely generalized at higher abstraction levels with more complex biometry description (e.g. behaviour recognition).

Therefore, the topics of this chapter, i.e., robust human body and face tracking, face recognition, and multisensor statistical fusion can represent a useful, although not complete, insight when building up a surveillance system capable of biometric functionalities. Let us now describe in more detail the content of the various papers:

Dockstader and Tekalp present a paper where the problem of modelling tracking failure is considered under a stochastic viewpoint. An interesting aspect of this paper is that it implicitly suggests that learning such a model, in addition to be of help in analysing scene sequences, can constitute by itself a biometric information. From an algorithmic viewpoint the key observation is the one of individuating a correlation between tracking failures and the error covariance matrix of a dynamic multiparameter Kalman-based shape model.

Marcialis and Roli present a multiple classifier approach to face recognition. The key contribution of this paper is the demonstration of the usefulness coming from a joint approach based on two well known statistical classifiers already used in the face recognition domain, namely Principal Component Analysis (PCA) and Linear Discriminant Analysis (LDA). The authors show that added value of fusion can be important, specially on databases where the variation of face poses, lighting conditions, etc, are high.

The problem of face tracking is considered in the paper by Loutas, Nikou, and Pitas. In this case, the goal of the paper is to provide a statistical model for face tracking. Used features are obtained from a set of high gradient points by applying Principal component analysis. The proposed probabilistic tracking method relies on both spatial grouping and temporal dynamic models. Robustness to both slight illumination and partial occlusions is effectively proven in the paper.

Finally, Osadciw, Varshney and Veeramachaneni present a paper oriented to define optimal fusion rules for multimodal biometric systems using multiple sensors. The paper provides an extended overview of biometric data acquisition modalities before using statistical multiple sensor fusion to design a careful analysis of the effects of sensor's performances on the selection of the final decision by a centralized classifier. The paper provides an insight in different possible fusion levels by exploring in more details decision level fusion. The necessity of scalability in multiple sensor systems for security as well as the possibility to choose different security system behaviours in presence of different operative contexts carries to the adaptive individuation of different optimal rules. To this end, a mathematical formulation of the problem is provided using statistical tools (e.g., Receiving Operating Characteristics, ROC curves). Such a formulation allows security system engineer to find different solutions with respect to desired false alarm and false detection rates.

Chapter 12

BIOMETRIC FEATURE EXTRACTION IN A MULTI-CAMERA SURVEILLANCE SYSTEM

Shiloh L. Dockstader and A. Murat Tekalp
Department of Electrical and Computer Engineering; University of Rochester; Rochester, NY 14627-0126 USA

Keywords: Human motion analysis, motion tracking, kinematic modeling, gait analysis, tracking failure prediction, Bayesian network, hidden Markov model

1. INTRODUCTION

Advanced, multi-camera surveillance systems present a practical and effective means of automatically detecting, tracking, and characterizing various people, events, activities, and situations. Recently, considerable focus has been placed on the tracking and analysis of human motion and, specifically, on the measurement of diverse biometrics for authentication and identification [1]. In this article we concentrate on the use of gait as a biometric and provide an introduction to a unique system for its robust tracking, extraction, and subsequent analysis.

1.1 Background

It is well known that a person's identity, actions, and intentions can often be inferred from a simple analysis of gait patterns. For instance, Cunado *et al.* [2] introduce a Hough transform technique for extracting frequency- and phase-based features from the lower body for use as potential biometrics. Similarly, BenAbdelkader *et al.* [3] present a constrained system for extracting metrics like cadence and stride length using assumptions of periodicity and constant velocity. Bobick and Johnson [4] establish a simple method for recognition and identification that recovers static body and gait parameters. In this chapter we present a straightforward means of biometric feature extraction that places a greater importance on initial tracking accuracy and robustness than on final recognition. As they have proven relevant in previous work, we concentrate on the measurement of variables like speed, cadence, stride length, and stance width and times.

Multiple-camera feature tracking is an obvious starting point for biometric measurement, as it affords the availability of 3-D data and an effective instrument for countering occlusion. Gavrila and Davis [5] present a vision system for the 3-D model-based tracking of unconstrained human movement. Using image sequences acquired simultaneously from multiple views, the authors recover 3-D body pose at each time instant without the use of markers. Stillman et al. [6] propose a robust system for tracking and recognizing multiple people with two cameras capable of panning, tilting, and zooming (PTZ) as well as two static cameras for general motion monitoring. Utsumi et al. [7] suggest a system for detecting and tracking multiple persons using multiple cameras to address complex occlusion and articulated motion. Their system is composed of multiple tasks including position detection, rotation angle detection, and body-side detection.

More sophisticated methods employ prior models of kinematic and dynamic structure for improved tracking accuracy. Kakadiaris and Metaxas [8], for example, use multiple views as well as hard kinematic constraints. For monocular 3-D body tracking, Sminchisescu and Triggs [9] suggest an improvement over classical sampling techniques that combines covariance scaling for local optimization with joint and non-self-intersection constraints. In contrast, Deutscher et al. [10] introduce an automatic partitioning of high dimensional spaces based on the application of a genetic-like crossover term to an annealed particle-filtering algorithm. Using activity-specific context, Sidenbladh et al. [11] and Pavlovi et al. [12] both introduce dynamic motion models for improved tracking, synthesis, and related tasks.

Finally, a rather critical concern in human motion understanding that has received very little attention is the analysis and prediction of terminal tracking failures. Pasqual et al. [13] describe an algorithm that explicitly addresses the uncertainty of tracking. They suggest a method of feature substitution using optical flow, texture, and implicit depth and then switch modalities as needed to prolong and enhance tracking performance. Darrell et al. [14] present an interesting approach to tracking using depth estimation, color segmentation, and intensity pattern classification. The method effectively increases tracking robustness using multiple modalities, but does not explicitly address the detection or prediction of tracking failures. Dockstader et al. [15] introduce a stochastic, geometric model for quantifying, but not predicting or detecting, tracking failures. For a more

12. Biometric feature extraction in a multi-camera surveillance system 221

thorough treatment of human motion analysis and related surveillance, we refer the reader to the excellent coverage provided in [16]-[18].

1.2 Contribution

In this chapter we present a multi-camera, video surveillance system that, unlike previous work, focuses exclusively on the difficulties of occlusion. We begin with an overview of Bayesian fusion for feature tracking and follow with a discussion on structural, stochastic kinematic, and dynamic modeling for biometric feature extraction. We complete the chapter with an introduction to tracking failure prediction for increased analysis accuracy and longevity. To implement this final innovation, we define a failure as an event and use a hidden Markov model to capture its temporal characteristics. This approach is applied to *each* parameter of a structural human body model and, in so doing, effectively enables the fault-tolerant tracking and extraction of motion-based biometric features. We demonstrate the effectiveness of our surveillance system on a number of complex video sequences, each exhibiting a substantial amount occlusion, clutter, and ambiguity.

2. SYSTEM CONFIGURATION AND THEORY

A high-level flow diagram of the suggested visual surveillance system is illustrated in Figure 12-1. The system comprises a number of components including structural modeling, 2-D observation generation, multi-camera fusion, 3-D kinematic and dynamic modeling, temporal integration, biometric feature extraction, and tracking failure prediction. For the sake of brevity we only provide an overview for many of these components, but refer the reader to more detailed treatments, where appropriate.

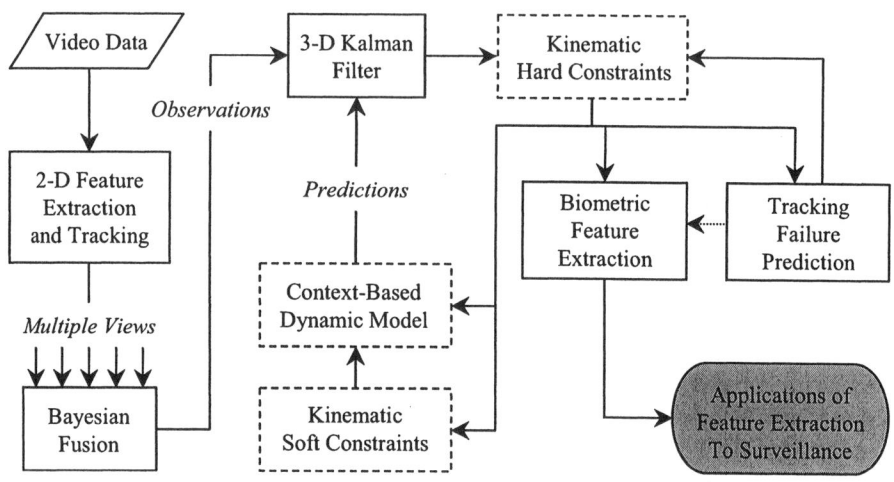

Figure 12-1. System flow diagram.

2.1 Multi-Camera Bayesian Feature Tracking

Let us assume there exists a group of 3-D feature points, possibly representing a collection of structural model parameters, that we wish to track as a function of time. The required 3-D data, $\mathbf{y}[k]$, at a particular frame, k, is not directly measurable but instead must be estimated from a set of 2-D image observations, $\{\xi_j[k]\}$, $1 \le j \le J$, taken from J views of a scene. Each measurement can be envisioned as the mean of a normal random variable and is accompanied by a corresponding noise covariance, $\mathbf{M}_j[k]$. If

$$\xi[k;K] \equiv \{\xi_{j_1}[k], \xi_{j_2}[k], \cdots, \xi_{j_K}[k]\} \subseteq [\xi_1[k], \xi_J[k]] \tag{12.1}$$

indicates a random set of K views upon which an estimate of 3-D position, $\hat{\mathbf{y}}[k]$, is based, then this estimation problem may be loosely formulated as one in which we calculate

$$(\hat{\mathbf{y}}, \hat{\xi}) = \arg\max_{\mathbf{y}, \xi} \{P_{\mathbf{Y}, \Xi}(\mathbf{y}, \xi)\}, \tag{12.2}$$

where, according to Bayes' rule,

$$P_{\mathbf{Y},\Xi}(\mathbf{y},\xi) = P_{\mathbf{Y}}(\mathbf{y}|\xi_J, \xi_{J-1}, \cdots, \xi_1) \cdot P_{\Xi}(\xi_J|\xi_{J-1}, \cdots, \xi_1) \cdots P_{\Xi}(\xi_1). \tag{12.3}$$

In (12.3), $P_{\mathbf{Y}}(\mathbf{y}|\xi_J, \xi_{J-1}, \cdots, \xi_1)$ models 3-D reconstruction noise, $P_{\Xi}(\xi_j)$ models 2-D observation noise, and the remaining conditional densities model the effects of occlusion and correlation *between* various views of the scene. Intuitively, these equations converge on the most likely observation subset, $\xi[k]$, and reconstructed value, $\hat{y}[k]$, given observed or calculated values of $\xi_j[k]$ and $\mathbf{M}_j[k]$ for the jth view as well as the inherent projective geometry between the J views.

This probabilistic formulation is easily and appropriately implemented as a Bayesian belief network. Specifically, we are given a number of inter-related variables that must be integrated in a topological fashion to maximize some context-dependent criterion. In our case, this is the consequent joint minimization of $\mathbf{R}[k]$, the 3-D noise covariance associated with $\hat{y}[k]$, and its J image plane projections. Within this architecture, each node of the Bayesian network represents a variable for a particular view; the only exception is the single child node that describes the reconstructed 3-D data.

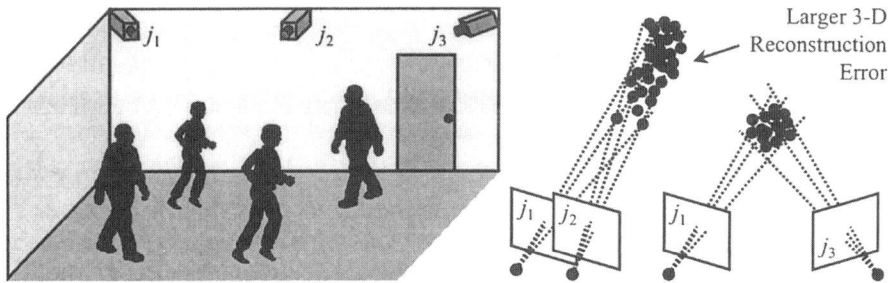

Figure 12-2. Camera configuration and its effect on 3-D reconstruction error.

To fully utilize the descriptive ability of the proposed Bayesian network and to model the causal dependencies between its variables, we order the nodes based on camera position as well as 2-D feature tracking confidence. As illustrated in Figure 12-2, for example, the placement of cameras within the scene introduces prior information regarding both the dependencies between the views as well as their ability to accurately reconstruct 3-D values. After establishing a topological node ordering, one proceeds by determining the appropriate network configuration that corroborates (12.3). The end result is a new set of estimates (3-D tracking observations), $\hat{y}[k]$, with an estimated noise covariance, $\mathbf{R}[k]$. For a more detailed description of this system component we refer the reader to [19].

2.2 Structural Body Modeling

The previously introduced observations, $\hat{y}[k]$, represent, in fact, the 3-D positions of various parameters associated with a structural model of the human body. We suggest a simple stick model defined by $N = 15$ parameters ($p_1...p_{15}$) that are measured in three-dimensional, body-centered coordinates, as indicated in the left of Figure 3-3. The origin of the coordinate system, p_0, corresponds to a fixed position on the 3-D stick model. The time-varying coordinate axes are uniquely determined at each frame, k, by interpreting the velocity of the origin. Points in this coordinate system are indicated as $\mathbf{z} = \{(x, y, z)\}$. We assume that during a normal gait cycle, the body moves forward, tangential to the transverse (*TP*, *x-y*) and sagittal (*SP*, *x-z*) planes and orthogonal to the coronal plane (*CP*, *y-z*). This is a reasonable assumption for typical surveillance systems in which gait is used as the primary biometric feature.

2.3 Kinematic and Dynamic Modeling

To limit the movement of our structural model and improve its tracking performance during periods of occlusion, we introduce a number of kinematic constraints. We place numerous hard constraints (i.e., absolute limits) on velocity, acceleration, joint angles, segment lengths, and parameter arrangements to encourage 3-D consistency. These restraints are applied iteratively during the final stage of 3-D parameter processing, immediately preceding biometric feature extraction and failure analysis as indicated in Figure 12-1. An example of a hard kinematic constraint is illustrated in the center of Figure 12-3 in which the relative distances between 3-D parameters is restricted using prior knowledge of human structure.

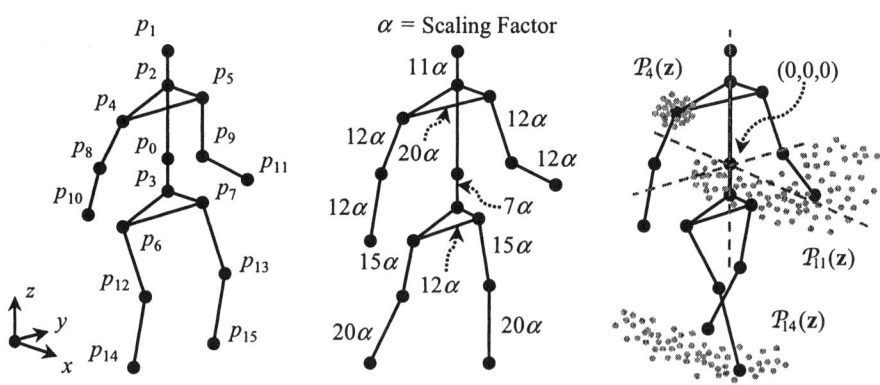

Figure 12-3. Structural body model (left) and absolute (center) and stochastic (right) kinematics.

In addition to hard constraints, we introduce the concept of stochastic kinematics, as illustrated in the right of Figure 12-3. Here, the mth structural parameter is associated with a prior density function, $P_m(\mathbf{z})$, represented as

12. Biometric feature extraction in a multi-camera surveillance system

a multi-dimensional point set, and constructed using previous examples of human body configuration exhibited during gait activities. The prior density is applied as an augmenting acceleration term within an existing 3-D dynamic motion model.

To track the structural model, we introduce a time-varying state vector, within the context of a Kalman filter, as $\sigma[k] \equiv [\sigma_1[k] \; \sigma_2[k] \; \cdots \; \sigma_{2N+1}[k]]^T$, where $\sigma_m[k]$, $m \leq N$ denotes the position of the mth parameter in our body-centered coordinate system and $\sigma_{m+N+1}[k] = \partial\sigma_m[k]/\partial k \big|_{m \leq N}$. For an estimate of the mth parameter we define a displacement model as

$$\hat{\sigma}_m[k \mid k-1] = \hat{\sigma}_m[k-1] + c_v \hat{\sigma}_{m+N+1}[k-1] \cdot \Delta t \\ + c_a (\Delta t)^2 (1 - l_m) \left(\frac{\omega_{m\hat{i}} - \hat{\sigma}_m[k-1]}{(\Delta t)^2} - \frac{c_v \hat{\sigma}_{m+N+1}[k-1]}{\Delta t} \right), \quad (12.4)$$

where, dropping the dependence of $\hat{i}_m[k]$ on m and k for simplicity,

$$\hat{i} \equiv \hat{i}_m[k] = \arg\min_i \{ \| \hat{\sigma}_m[k-1] - \omega_{mi} \| \}, \quad (12.5)$$

$$l_m = \frac{P_m(\hat{\sigma}_m[k-1])}{P_m(\omega_{m\hat{i}})} \in [0,1], \quad (12.6)$$

$$\omega_m \equiv \{\omega_{m1} \; \cdots \; \omega_{m|\omega_m|}\} = \{\mathbf{z} \mid \nabla P_m(\mathbf{z}) = \mathbf{0}, \; |HP_m(\mathbf{z})| < 0\}, \quad (12.7)$$

and Δt is the time sampling between frames, c_v and c_a are kinematic constants for velocity and acceleration, and H indicates the Hessian of a real-valued function.

The above dynamics are easily integrated into the framework of a Kalman filter with an existing motion model. Assuming this is the case, for an individual parameter we have

$$\hat{\sigma}_m[k] = \hat{\sigma}_m[k \mid k-1] + \mathbf{D}_m[k]\left(\hat{\mathbf{y}}_m[k] - \mathbf{\Phi}_m[k]\hat{\sigma}_m[k \mid k-1]\right), \quad (12.8)$$

$$\hat{\sigma}_{m+N+1}[k] = \tfrac{1}{\Delta t}\{\hat{\sigma}_m[k] - \hat{\sigma}_m[k-1]\}, \quad (12.9)$$

and

$$\mathbf{\Gamma}_m[k] = (\mathbf{I} - \mathbf{D}_m[k]\mathbf{\Phi}_m[k])\mathbf{\Gamma}_m[k \mid k-1]. \quad (12.10)$$

Here, $\mathbf{D}_m[k]$ is the Kalman gain matrix, $\mathbf{\Phi}_m[k]$ is the linear observation matrix, $\mathbf{\Gamma}_m[k\,|\,k-1]$ is the predicted mean-square error (MSE) matrix, and $\hat{\mathbf{y}}_m[k]$ is the given 3-D observation. Note that the state update is subject to the application of hard kinematic constraints. We describe this process as well as the intermediate filtering steps in greater detail in [15] and [19].

2.4 Biometric Feature Extraction

We extract a set of motion-based biometric features, $\boldsymbol{\pi}_S[k]$, from the structural parameters. Given a sufficient structural model and an effective tracking paradigm, feature extraction becomes a relatively simple task. Omitting the dependence of the variables on k for the sake of notational simplicity, we have for gait velocity [speed] $\pi_{S0} = \|\hat{\boldsymbol{\sigma}}_{N+1}\|$. We estimate stride length as $\pi_{S1} = (\hat{\boldsymbol{\sigma}}_{14} - \hat{\boldsymbol{\sigma}}_{15}) \cdot \hat{x}$ and stance width as

$$\pi_{S2} = [\delta_L \hat{\boldsymbol{\sigma}}_{15} - (1-\delta_L)\hat{\boldsymbol{\sigma}}_{13} - \delta_R \hat{\boldsymbol{\sigma}}_{14} - (1-\delta_R)\hat{\boldsymbol{\sigma}}_{12}] \cdot \hat{y}, \qquad (12.11)$$

where $\delta_L = |\mathbf{\Gamma}_{13}||\mathbf{\Gamma}_{15}|^{-1}$ and $\delta_R = |\mathbf{\Gamma}_{12}||\mathbf{\Gamma}_{14}|^{-1}$. We indicate right and left stance times as $\pi_{S3} = u(\|\hat{\boldsymbol{\sigma}}_{N+15}\| - \varepsilon_V)$ and $\pi_{S4} = u(\|\hat{\boldsymbol{\sigma}}_{N+16}\| - \varepsilon_V)$, where $u(\cdot)$ represents a unit-step function while ε_V denotes an error-minimizing velocity threshold. We approximate arm swing with $\pi_{S5} = \|\hat{\boldsymbol{\sigma}}_{11} - \hat{\boldsymbol{\sigma}}_{10}\|$ and cadence, measured in footsteps per minute, with

$$\pi_{S6}[k] = \frac{30}{\Delta t(k+1)} \sum_{j=0}^{k} \left[\left|\frac{\partial \pi_{S3}[i]}{\partial i}\right|_{i=j} + \left|\frac{\partial \pi_{S4}[i]}{\partial i}\right|_{i=j} \right]. \qquad (12.13)$$

One might also estimate double stance time, an important variable for event analysis, with $\pi_{S7} = \pi_{S3} + \pi_{S4}$. For an illustration of some of these biometric features, we refer the reader to Figure 12-4.

2.5 Tracking Failure Prediction

A terminal tracking failure is defined as an immediate and sustained loss in tracking accuracy at a specific structural [model] parameter. We quantify an immediate loss (p'_m) via the distance between a parameter's estimated and ground-truth values and a sustained loss (p''_m) via the RMS error between the same measures taken over a period of T frames.

12. Biometric feature extraction in a multi-camera surveillance system

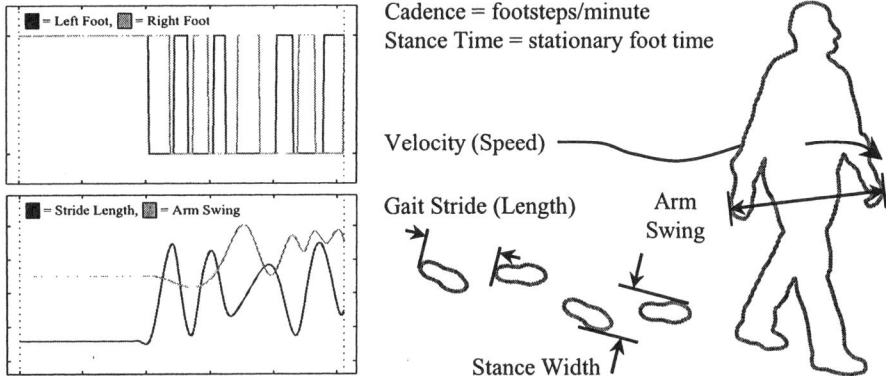

Figure 12-4. Motion-based biometric features and gait variables.

An analysis of typical image sequences used for tracking and interpreting human motion shows that most failures are preceded by $T \equiv 30$ or fewer frames of relevant data. Although tracking failures are partially dependent on the nature of the underlying motion, they typically develop quickly and exhibit similar degradations in key features regardless of the particular activity or event in progress. The error thresholds are known to vary as a function of both the average error and standard deviation of the mth structural parameter, as reported in [15], and are summarized in Figure 12-5.

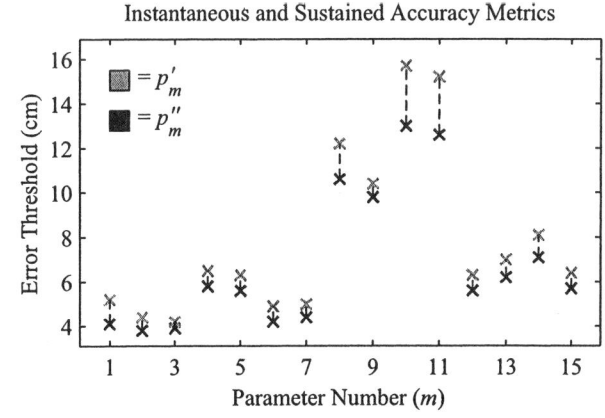

Figure 12-5. Empirically derived tracking failure threshold parameters.

Tracking failures are not easily characterized by the changing positions of structural parameters over time, but are correlated, however, with temporal changes in noise covariance measurements. Let an observation derived from the mth parameter at frame k be denoted by the determinant of its MSE matrix, $o_m[k] = |\Gamma_m[k]|$, and let

$$\mathbf{o}_k \equiv [o_1[k] \; o_2[k] \; \cdots \; o_N[k]]^T \qquad (12.13)$$

indicate the observation vector actually used by each parameter for failure analysis. A single observation sequence is denoted by $\mathbf{O} = (\mathbf{o}_1 \; \mathbf{o}_2 \; \cdots \; \mathbf{o}_T)$.

To model failures in a stochastic way, we introduce a nearly ergodic hidden Markov model (HMM), $\lambda = (A, B, \zeta)$, for each of the N structural parameters. Using video sequences with an observation length of $T = 30$ and vectors of dimension $N = 15$, one finds in practice that the observations are conveniently clustered, independently for each parameter, into $M = 64$ discrete symbols. The number of states, R, used in each model is motivated by the nature of tracking failures. In typical sequences, failures show a strong dependency on the confidence of directly connected parameters, but a much weaker dependency on all others. For most structural parameters, the empirically derived value of $R = 6$ produces a reasonably accurate description of this phenomenon. A generic Markov topology is illustrated in Figure 12-6.

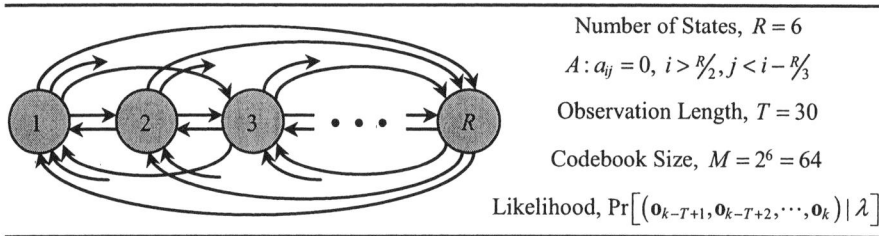

Figure 12-6. Hidden Markov model topology for failure event detection.

Estimating the remaining parameters of λ, where λ loosely refers to any of the Markov models, is performed using the Baum-Welch method [20]. The only restriction on the topology of the Markov models is that for the state-transition probabilities, where $A = \{a_{ij}\}$,

$$a_{ij} = 0, \; i > R/2, \; j < i - R/3. \qquad (12.14)$$

This constraint allows for the progression of noise covariance changes from which the algorithm cannot recover (e.g., a tracking degradation of multiple connected parameters). Any model is then estimated according to

12. Biometric feature extraction in a multi-camera surveillance system

$$\hat{\lambda} = \arg\max_{\lambda}\left(\prod_i \Pr[\mathbf{O}^{(i)}|\lambda]\right), \quad (12.15)$$

where $\mathbf{O}^{(i)}$ is the ith observation training sequence. The optimization procedure supports the initial choices of R, M, and T by producing local maxima at the chosen values. For an arbitrary Markov model, the initial state distribution, ζ, the state-transition probabilities, A, and the observation symbol probabilities, B, are then calculated accordingly. Initial values for these parameter estimates are based on a uniform distribution.

For all sequences introduced after the training set and for any structural parameter, the metric $\Pr\left[(\mathbf{o}_{k-T+1},\cdots,\mathbf{o}_k)|\lambda\right]$ or, alternatively, $\Pr[\mathbf{O}|\lambda]$ may be used to test the likelihood that the observation, in which $T=30$, was produced by λ. A greater likelihood implies a higher correlation between the observed sequence and those known to correspond with imminent tracking failures. A simple threshold placed on this output probability is a sufficient test for detecting such events. Thus, where a single fixed-state sequence is denoted by $\mathbf{r}=(r_1\ r_2\ \cdots\ r_T)$, we have

$$\Pr[\mathbf{O}|\lambda] \equiv \sum_{\text{all }\mathbf{r}} \Pr[\mathbf{O}|\mathbf{r},\lambda]\cdot\Pr[\mathbf{r}|\lambda] \overset{\text{Failure}}{\underset{\text{No Failure}}{\gtrless}} \lambda'. \quad (12.16)$$

The above output probability can be estimated using the well-known forward estimation procedure [20].

Figure 12-7. Accurately tracked and extracted frames from multiple views.

An analysis of the above output probability, taken as a function of frame number, reveals a nearly logarithmic relationship for $\Pr[\mathbf{O}|\lambda] \gg 0$. This suggests that each HMM could be used for prediction as well as detection. Thus, we introduce a parametric mapping function between the output probability, $\Pr[\mathbf{O}|\lambda]$, and the expected number of frames until the next tracking failure, k_f, according to

$$k_f = a_0 + a_1 \ln\left(\Pr[\mathbf{O}|\lambda] + a_2\right). \tag{12.17}$$

The parameters (a_0, a_1, a_2) are solved using non-linear least squares, implemented via the Levenberg-Marquardt approach. Each structural parameter is assigned a unique set of logarithmic parameters to handle its specific relationship between frames and probability.

The proposed failure prediction is used in a recursive fashion to prevent the enforcement of any kinematic constraints involving multiple parameters in which one or more is expected to fail within $k'_f = T/4$ frames. Naturally, this failure mechanism is used to prevent the erroneous extraction of gait-based biometric features as well. These aspects are illustrated in Figure 12-1. The end result due to failure prediction is a considerable increase in tracking accuracy and longevity, especially in the presence of occlusion.

3. EXPERIMENTAL RESULTS AND ANALYSIS

We implement the proposed surveillance system using a network of three or four calibrated, synchronized cameras and validate the algorithms with approximately 75 minutes of video collected at a rate of $\Delta t = \frac{1}{30}\sec$. Nearly 70% of the data is used to construct and train our mathematical and stochastic models; the remainder is reserved for testing and analysis. Using this data, we are able to track and extract biometric features for an average of 132 frames between structural model parameter failures. This is notably better performance than the 37 frames obtained without the use of our Bayesian fusion, kinematic modeling, and recursive failure analysis components. Frames extracted from the video sequence from multiple views are shown in Figure 12-7.

12. Biometric feature extraction in a multi-camera surveillance system

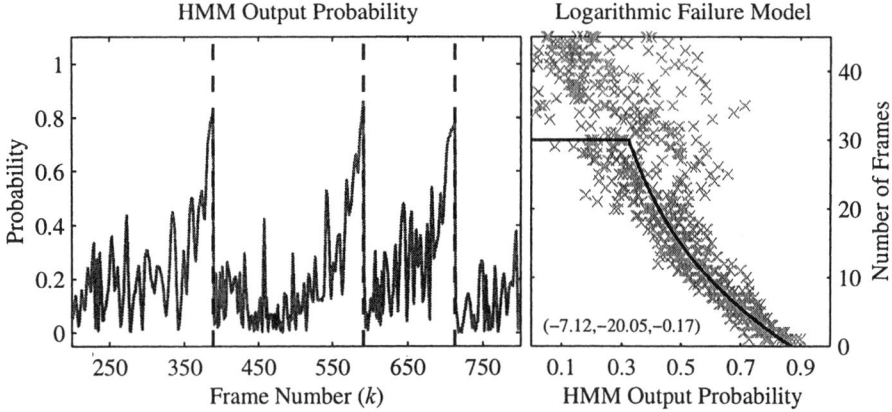

Figure 12-8. HMM output probability (left) and logarithmic failure prediction (right).

The addition of tracking failure analysis truly enables the utility of the suggested surveillance system. In Figure 12-8, we illustrate on the left the output probability, $\Pr[\mathbf{O}|\lambda]$, for an individual structural parameter, p_{10}; the vertical dashed lines indicate actual tracking failures. It is immediately evident from the figure that there exists a causal relationship, as hypothesized, between tracking failures and the proposed hidden Markov model output probability. Using a specificity-maximizing threshold for failure detection, one finds an average accuracy of 97.2% and an average sensitivity of 87.2% across all structural parameters.

The predictive capability of the logarithmic model is demonstrated for p_{10} in the right of Figure 12-8. We show the estimated model as well as a subset of observation pairs. Although the prediction variance increases considerably for $k_f \gg 1$, it remains relatively bounded for $\Pr[\mathbf{O}|\lambda] \gg 0$, thus maintaining an ability to reasonably predict failures for low values of k_f (e.g., $0 < k_f < T/2$). The logarithmic prediction model for p_{10}, for example, exhibits fairly accurate behavior. A failure predicted to occur within 1-2 frames actually occurs within 1-2 frames nearly 60% of the time, 3-4 frames nearly 20% of the time, and 5-6 frames approximately 8% of the time.

To further quantify the performance of the failure prediction and, ultimately, the Markov hypothesis, we compare $\Pr[\mathbf{O}|\lambda]$ to a more fundamental metric, $o_m[k]$. This secondary measure considers neither the temporal correlations inherent in a failure event, nor the parameter dependencies within the structural body model. For $o_m[k]$, we calculate an average accuracy of 91% and an average sensitivity of only 43%.

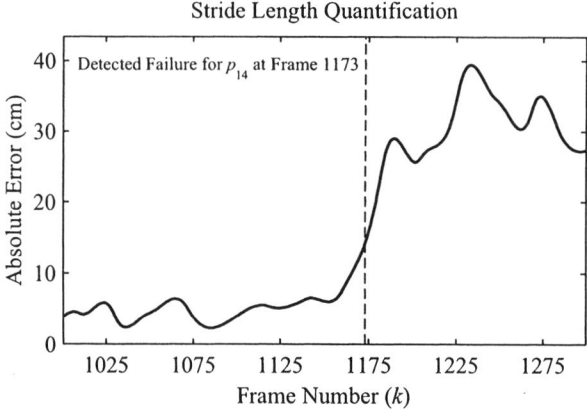

Figure 12-9. Stride length error before and after a detected tracking failure.

Here, the cost of maintaining a high specificity is an elevated rate of incorrectly detected tracking failures, thus leading to an incomplete biometric feature extraction. In the case of feature extraction, the value of the proposed failure prediction is especially clear. In Figure 12-9, we illustrate the error in calculated stride length before and after a correctly detected failure at parameter p_{14}. Using failure prediction the extraction is terminated at the estimated failure point. Without it, however, an erroneous biometric is established that, in turn, has the potential of presenting a serious, negative impact to any subsequent analysis or interpretation.

4. CONCLUSIONS

This article presents an original and effective multi-camera video surveillance system for robust biometric feature extraction. We begin with a method for generating 3-D observations using an occlusion-adaptive, Bayesian belief network. Following is a discussion of an appropriate human body model limited by a variety of constraints, including a new class of probabilistic kinematics. We close with the extraction of biometric features and the introduction of a new approach to tracking failure prediction. The latter novel concept defines a failure as an event and characterizes its temporal nature using a hidden Markov model. The proposed feature extraction system is found to be exceptionally effective in the presence of complex human motion and, in particular, severe occlusion.

ACKNOWLEDGEMENTS

This research was supported in part by grants from the Center for Future Health, CEIS, and Eastman Kodak Company. The authors also thank Dr. Michel J. Berg for his many insightful comments and suggestions and Nikita S. Imennov for his assistance with data collection and analysis.

REFERENCES

[1] Diechmann U., Plankensteiner P., Schamburger R., Froba B., and Meller S. SESAM: A biometric person identification system using sensor fusion. *Proc of the Int Conf on Audio and Video-Based Biometric Person Authentication.* 12-14 Mar 1997; Crans-Montana, Switzerland; pp. 301-310.

[2] McKenna S.J. and Gong S. Non-intrusive person authentication for access control via visual tracking and face recognition. *Proc of the Int Conf on Audio and Video-Based Biometric Person Authentication.* 12-14 Mar 1997; Crans-Montana, Switzerland; pp. 177-183.

[3] Cunado D., Nixon M.S., and Carter J.N. Using gait as a biometric, via phase-weighted magnitude spectra. *Proc of the Int Conf on Audio and Video-Based Biometric Person Authentication.* 12-14 Mar 1997; Crans-Montana, Switzerland; pp. 95-102.

[4] BenAbdelkader C., Cutler R., and Davis L. Stride and cadence as a biometric in person identification and verification. *Proc of the Int Conf on Automatic Face and Gesture Recognition.* 20-21 May 2002; Washington, DC; pp. 372-377.

[5] Bobick A.F. and Johnson A.Y. Gait recognition using static, activity-specific parameters. *Proc of the Conf on Computer Vision and Pattern Recognition.* 8-14 Dec 2001; Kauai, HI; 1, pp. 423-430.

[6] Gavrila D.M. and Davis L.S. Model-based tracking of humans in action: A multi-view approach. *Proc of the Conf on Computer Vision and Pattern Recognition.* 18-20 Jun 1996; San Francisco, CA; pp. 73-80.

[7] Stillman S., Tanawongsuwan R., and Essa I. A system for tracking and recognizing multiple people with multiple cameras. *Proc of the Int Conf on Audio and Video-Based Biometric Person Authentication.* 22-23 Mar 1999; Washington, DC; pp. 96-101.

[8] Utsumi A., Mori H., Ohya J., and Yachida M. Multiple-human tracking using multiple cameras. *Proc of the Int Conf on Automatic Face and Gesture Recognition.* 14-16 Apr 1998; Nara, Japan; pp. 498-503.

[9] Kakadiaris I.A. and Metaxas D. Model-based estimation of 3D human motion with occlusion based on active multi-viewpoint selection. *Proc of the Conf on Computer Vision and Pattern Recognition.* 18-20 Jun 1996; San Francisco, CA; pp. 81-87.

[10] Sminchisescu C. and Triggs B. Covariance scaled sampling for monocular 3D body tracking. *Proc of the Conf on Computer Vision and Pattern Recognition*. 8-14 Dec 2001; Kauai, HI; 1, pp. 447-454.

[11] Deutscher J., Davison A., and Reid I. Automatic partitioning of high dimensional search spaces with articulated body motion capture. *Proc of the Conf on Computer Vision and Pattern Recognition*. 8-14 Dec 2001; Kauai, HI; 2, pp. 669-676.

[12] Sidenbladh H., Black M.J., and Sigal L. Implicit probabilistic models of human motion for synthesis and tracking. *Proc of the Eur Conf on Computer Vision*. 28-31 May 2002; Copenhagen, Denmark; 2350, pp. 784-800.

[13] Pavlovi V., Rehg J.M., Cham T.-J., and Murphy K. A dynamic Bayesian network approach to figure tracking using learned dynamic models. *Proc of the Int Conf on Computer Vision*. 20-27 Sep 1999; Kerkyra, Greece; 1, pp. 94-101.

[14] Pasqual A.A., Aizawa K., and Hatori M. Use of multiple visual features for object tracking. *Proc of SPIE*. 25-27 Jan 1999; San Jose, CA; 3653, pp. 946-955.

[15] Darrell T., Gordon G., Harville M., and Woodfill J. Integrated Person Tracking Using Stereo, Color, and Pattern Detection. *Int J of Computer Vision*. 2000; 37(2), pp.175-185.

[16] Dockstader S.L., Berg M.J., and Tekalp A.M. Stochastic Kinematic Modeling and Feature Extraction for Gait Analysis. to appear in *IEEE Trans on Image Processing*. 2003.

[17] Gavrila D.M. The Visual Analysis of Human Movement: A Survey. *Computer Vision and Image Understanding*. 1999; 73(1), pp. 82-98.

[18] Aggarwal J.K. and Cai Q. Human Motion Analysis: A Review. *Computer Vision and Image Understanding*. 1999; 73(3), pp. 428-440.

[19] Collins R.T., Lipton A.J., and Kanade T. Introduction to the Special Section on Video Surveillance. *IEEE Trans on Pattern Analysis and Machine Intelligence*. 2000; 22(8), pp. 745-746.

[20] Dockstader S.L. and Tekalp A.M. Multiple Camera Tracking of Interacting and Occluded Human Motion. *Proc of the IEEE*. 2001; 89(10), pp. 1441-1455.

[21] Rabiner L. and Juang B.-H. *Fundamentals of Speech Recognition*, Upper Saddle River, NJ: Prentice Hall, 1993.

Chapter 13

FUSION OF FACE RECOGNITION ALGORITHMS FOR VIDEO-BASED SURVEILLANCE SYSTEMS

Gian Luca Marcialis and Fabio Roli
Department of Electrical and Electronic Engineering - University of Cagliari – Italy

Keywords: Video Surveillance, Biometrics, Face Recognition, Fusion of Multiple Classifiers .

1. INTRODUCTION

It is widely acknowledged that face recognition could play an important role in advanced video-based surveillance systems, mainly because it is non intrusive and does not require people cooperation [1-2]. Unfortunately, face recognition algorithms showed to suffer a lot from the high variability of environmental conditions. As an example, the effectiveness of face recognition strongly depends on lighting conditions and on variations in the subject's pose and expression in front of the camera. This obviously limits their application to real video-surveillance tasks. On the other hand, face is considered a very good biometric. People recognize each other through the face, the acquisition process is non-intrusive, and does not require the collaboration of the subject to be recognized. Therefore, face recognition is a very active research field with many applications. For the purposes of this chapter, the face recognition applications can be subdivided in two types: applications in controlled and uncontrolled environments. One of the main applications of the first type is the so called "identity authentication". A person submits to the automatic identity verification system its face (frontal and/or profile view) and declares her/his identity. The system matches the acquired face with the "template" stored in its data base, and classifies the person as a "genuine" (i.e., the claimed identity is accepted) or an "impostor". Automatic identity verification based on face recognition is usually performed in controlled environments, and requires person cooperation.

Applications of the second type refer to the problem of recognition of an identity in a scene, and they are very useful for video-surveillance tasks. The recognition system first detects the face in the image and normalizes it with respect to the pose, lighting, and scale conditions. Then, it tries to associate the face to one or more faces stored in its database, and provides as outputs the set of faces that are considered as "nearest" to the detected face. This problem is much more complex than the previous "verification" problem. It is computationally expensive and needs of robust algorithms for detection, normalization, and recognition. In the context of video surveillance applications, the following problems can seriously affect face recognition performances:

- the scene complexity, that can strongly affect face detection performances [3];
- the quality of video sequence, that can be very low due to the poor performance of the surveillance cameras, and the very large variations of lighting conditions;
- the small size of acquired faces;
- the very large variations of face expression and pose.

Usually, each of the above problems is so complex that it must be addressed separately. In this chapter, we focus on the last stage of the face recognition process. We assume that the previous steps of face detection, restoration, and normalization have been already performed. In section 2, the state of the art of face recognition approaches is briefly reviewed, and the novel role of algorithm fusion is pointed out. In section 3, fusion of two well-known face recognition algorithms, namely, PCA and LDA, is proposed. In section 4, experimental results are reported. Conclusions are drawn in section 5.

2. FACE RECOGNITION SYSTEMS: A SHORT OVERVIEW

A good survey of the state-of-the-art of face recognition systems can be found in [4]. In the following, we briefly review the main works.

Many face recognition systems have been proposed in the last years. Each of them is based on a particular representation of face. For the purposes of this Chapter, we can identify two main types of approaches: the so called appearance-based approaches, where a feature vector for characterizing the face is derived from the input image, and the structural approaches, where a deformable model, like a graph, is used for face representation.

The term "appearance-based" has been proposed for distinguishing the statistical approaches from the structural ones [1-2]. The appearance-based methods describe the face with a feature vector derived from the original input image. The feature vector is computed by reducing the dimensionality of the original image space. Feature reduction is performed by applying some standard pattern recognition algorithms. The aim is to reduce the redundant and/or noisy information contained in the original image. Consequently, a compact and effective description of the face image is obtained.

The most used approach is the face representation by Principal Component Analysis (PCA), or "eigenface" approach, proposed by Turk and Pentland [5]. The face image is projected to a space where the correlation among the features is zero. Only the components with highest variance are used for characterizing the face. A transformation that satisfies this condition is the so-called Karhunen-Loeve transform. Another appearance-based approach is the face representation by Linear Discriminant Analysis (LDA), or "fisherface" approach, proposed by Kriegmann et al. [6]. The face image is projected to the so called Fisher space, in which the variability among the face-vectors of the same class is minimized, and the variability among the face-vectors of different classes is maximized. In this case, the face is represented by a number of components smaller than the one of the PCA. We discuss in more detail both "eigenface" and "fisherface" approaches in section 3. Usually, the matching between two face feature vectors is performed by applying some kind of metric like the Euclidean distance, the Mahalanobis distance, etc.

The Local Feature Analysis (LFA) by Penev and Atick [7] derives from the analysis of the local information around some critical points of the face (e.g., eyes, nose, lips). This local information can be computed through a kernel function centered on the given critical points. An example of such kernel function is given by the PCA transform. LFA is the face representation algorithm used in the face recognition system developed by Identix company.

With regard to the structural approaches, a well-known algorithm is the so called elastic bunch graph method [8] that refers to the dynamic link architectures [9] proposed by Wiskott et al. A set of reference points, called "fiducial points", is selected in the face image. Each fiducial point is a node of a fully connected graph, and it is labelled by the Gabor filters responses computed in a window centered around the fiducial point. Each arch is labelled with the distance between the correspondent fiducial points. Recognition is performed by an elastic matching between two graphs.

It is worth noting that, from the viewpoint of video-surveillance applications using video sequences, the above algorithms can be applied within the so called "still-to-still" and "multiple-stills-to-still" face recognition paradigms. In the still-to-still paradigm, the recognition algorithm (e.g., the Local Feature Analysis) is applied only if a good "pose" (e.g., a frontal view of the person to be recognized) can be detected in the video sequence. Therefore, the video sequence is firstly processed in order to detect a frame associated to a good pose (usually, a frontal view). Then, the recognition algorithm is applied to such frame. This approach requires a good pose estimation algorithm. In the multiple-stills-to-still approach, templates associated to multiple poses and expressions are used to cover all possible variations of the face in the video sequence. Therefore, recognition can be attempted for the most of frames of video sequence. The problem of this approach is how to choose the most representative face poses, because of the very large cases to be handled.

In the above paradigms, no temporal information and correlation among images is used.

Recently, Krüger, Zhou [10] and Chellappa [11] proposed the "video-to-video" paradigm, where the whole sequence of faces acquired during a given time interval of the video sequence is associated to a class (identity). This concept implies the temporal analysis of the video sequence with dynamical models (e.g., Bayesian models), and the "condensation" of the tracking and recognition problems. These methods are a matter of on-going research, and the reported experiments were performed without "real" variations of pose and face expressions.

Other face recognition systems based on the still-to-still and multiple-stills-to-still paradigms have been proposed [12-13]. However, none of them is able to effectively handle the large variability of critical parameters, like pose, lighting, scale, face expression, some kind of forgery in the subject appearance (e.g., the beard). Effective handling of lighting, pose and scale variations is a matter of on-going research. Typically, a face recognition system is specialized on a certain type of face view (e.g. frontal views), disregarding the images that do not correspond to such view. Therefore, a powerful pose estimation algorithm is required. But this is often not sufficient, and an unknown pose can deceive the whole system. Therefore, a face recognition system can usually achieve good performance only at the expense of robustness and reliability.

In order to improve the performance and robustness of individual recognizers, the use of multiple classifier systems (MCSs) has been recently proposed. MCSs are currently a very active research field [14]. Multiple classifiers systems cover a wide spectrum of applications: handwritten character recognition, fingerprint classification and matching, remote-sensing images classification, etc. The effectiveness of this approach is documented by many experimental results [14].

Approaches for improving the performance and the robustness of face recognition using MCSs have been proposed. Achermann and Bunke [15] proposed the fusion of three recognizers based on frontal and profile faces. The outcome of each expert, represented by a score, i.e., a level of confidence about the decision, is combined with simple fusion rules (majority voting, rank sum, Bayes's combination rule). Lucas [16] used a n-tuple classifier for combining the decisions of experts based on sub-sampled images. Tolba [17] presented a simple combination rule for fusing the decisions of RBF and LVQ networkz. Marcialis and Roli [18-19] reported preliminary experiments on the fusion of two statistical approaches, PCA and LDA, for face verification and recognition.

3. FACE RECOGNITION BY FUSION OF STATISTICAL FACE REPRESENTATIONS

In this section, we present our methodology for fusing two appearance-based (or statistical) approaches to face recognition: the PCA representation ("eigenface" approach) and the LDA representation ("fisherface" approach). We already used the fusion of LDA and PCA for face verification with good results [19]. From the viewpoint of video surveillance applications, it is worth noting that our methodology should be applied according to the still-to-still paradigm (Section 2). Figure 13-1 gives an overview of the proposed method. It is implemented by the following steps:
- representation of the face image according to the PCA and the LDA approaches;
- the distance vectors d^{PCA} and d^{LDA} of the input image from all the N face templates stored in the database are computed;
- for the final decision, these two vectors are fused by a combination rule. We proposed two algorithms for the fusion phase: the K-Nearest Neighbors and the Nearest Mean.

In the following, we briefly describe the theoretical framework behind the two face representations.

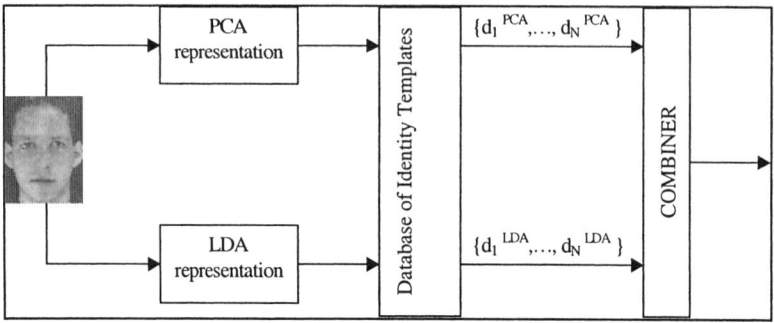

Figure 13-1. The flow diagram of our Face Recognition System

3.1 PCA and LDA representations for Face Recognition

Let X be a *d*-dimensional feature vector. In our case, *d* is equal to the number of pixel of each face image. The high dimensionality of the related "image space" is a well-known problem for the design of a good face recognition algorithm. Therefore, methods for reducing the dimensionality of such image space are required. To this end, Principal Component Analysis (PCA) and Linear Discriminant Analysis (LDA) are widely used.

Principal Component Analysis [5, 20] is defined by the transformation:

$$y_i = W^t x_i \qquad (13.1)$$

Where $x_i \in X \subseteq \Re^d$, $i = 1,...,n$ (*n* samples). *W* is a *d*-dimensional transformation matrix whose columns are the eigenvectors related to the eigenvalues computed according to the formula:

$$\lambda e_i = S e_i \qquad (13.2)$$

S is the scatter matrix (i.e., the covariance matrix):

$$S = \sum_{i=1}^{n}(x_i - m) \cdot (x_i - m)^t, \qquad m = \frac{1}{n}\sum_{i=1}^{n} x_i \qquad (13.3)$$

This transformation is called Karuhnen-Loeve transform. It defines the *d*-dimensional space in which the covariance among the components is zero, because the covariance matrix is diagonal. The eigenvalues correspond to the variances of each component in the transformed space. After ordering the eigenvalues by increasing order, it is possible to consider a small number of "principal" components exhibiting the highest variance. The principal components of the transformed space are also called the most expressive features, and the eigenvectors related to the most expressive features are called "eigenfaces".

The Linear Discriminant Analysis (also called Fisher Discriminant Analysis) [6, 20] is defined by the transformation:

$$y_i = W^t x_i \tag{13.4}$$

The columns of W are the eigenvectors of $S_w^{-1} S_b$, where S_w is the *within-class scatter matrix*, and S_b is the *between-class scatter matrix*. It is possible to show that this choice maximizes the ratio $\det(S_b)/\det(S_w)$.

These matrices are computed as follows:

$$S_w = \sum_{j=1}^{c} \sum_{i=1}^{n_j} (x_i^j - m_j) \cdot (x_i^j - m_j)^t, \quad m_j = \frac{1}{n_j} \sum_{i=1}^{n_j} x_i^j \tag{13.5}$$

Where x_i^j is the *i*-th pattern of *j*-th class, and n_j is the number of patterns for the *j*-th class.

$$S_b = \sum_{j=1}^{c} (m_j - m) \cdot (m_j - m)^t, \quad m = \frac{1}{n} \sum_{i=1}^{n} x_i \tag{13.6}$$

The eigenvectors of LDA are called "fisherfaces". LDA transformation is strongly dependent on the number of classes (*c*), the number of samples (*n*), and the original space dimensionality (*d*). It is possible to show that there are almost *c*-1 nonzero eigenvectors. *c*-1 being the upper bound of the discriminant space dimensionality. We need *d+c* samples at least to have a nonsingular S_w. It is impossible to guarantee this condition in real applications. Consequently, an intermediate transformation is applied to reduce the dimensionality of the image space. To this end, we used the PCA transform [21]. Other regularization techniques have been proposed [22-26].

3.2 Fusion of PCA and LDA for Face Recognition

Many works analyzed the differences between these two techniques (see, in particular, [6]), but no work investigated the possibility of fusing them. In our opinion, the apparent strong correlation of LDA and PCA, especially when frontal views are used and PCA is applied before LDA, discouraged the fusion of such algorithms. However, it should be noted that LDA and PCA are not so correlated as one can think, as the LDA transformation applied to the principal components can generate a feature space significantly different from the PCA one. Therefore, the fusion of LDA and PCA for face recognition and verification is worth of theoretical and experimental investigation.

We propose two kind of approaches to fuse PCA and LDA face representations: the K-Nearest Neighbor approach (KNN) and the Nearest Mean approach (NM) [20].

First of all, we normalize the distance vectors d^{PCA} and d^{LDA} in order to map the range of these distances to the interval [0,1]:

$$d_{norm} = \frac{d - d_{min}}{d_{max} - d_{min}} \qquad (13.7)$$

Then, a *combined distance vector d* that must contain both PCA and LDA informations is computed. To this end, the following two techniques can be used:
- the combined distance vector is computed as the mean vector:

$$d = \left\{ \frac{d_1^{PCA} + d_1^{LDA}}{2}, ..., \frac{d_N^{PCA} + d_N^{LDA}}{2} \right\} \qquad (13.8)$$

- the combined distance vector is computed by appending the d^{PCA} and d^{LDA} vectors:

$$d = \left\{ d_1^{PCA}, ..., d_N^{PCA}, d_1^{LDA}, ..., d_N^{LDA} \right\} \qquad (13.9)$$

where N is the number of images in the database. If C is the number of the identities, also called *classes*, an identity c is associated to each couple $\left(d_j^{LDA}, d_j^{PCA} \right)$, $j = 1,...,N$.

After computing and ordering the combined distance vector d, we follow the KNN strategy: *the most frequent identity among the first K components of d is selected*. If the combined distance vector follows eq. (13.8), we call our algorithm "Mean-KNN" (M-KNN); if it follows eq. (13.9), we call our algorithm "Append-KNN" (A-KNN).

In the case of the NM approach, we first compute a template for each identity in the database. We selected the average image for both PCA and LDA representations. Consequently, our distance vectors d^{PCA} and d^{LDA} are composed by C components instead of N. These vectors are combined according to eq. (13.8) or (13.9). The identity associated to the smallest combined distance is selected. The related algorithms are called "Mean-NM" (M-NM), and "Append-NM" (A-NM), respectively.

4. EXPERIMENTAL RESULTS

In this section, we report our experiments on two well-known face data bases: the AT&T and the Yale datasets.

4.1 Data Sets

The AT&T data set is made up of ten different images of 40 distinct subjects. For some subjects, the images were taken at different times, varying the lighting, facial expressions (open/closed eyes, smiling/not smiling), and facial details (glasses/no glasses). All the images were taken against a dark homogeneous background with the subjects in an upright, frontal position (with tolerance for some side movement). The data set was subdivided into a training set, made up of 5 images per class/identity (200 images), and a test set, made up of 5 images per class (200 images). In order to assess recognition performances, we repeated our experiment for ten random partitions of the data set. Reported results refer to the average performance of such ten runs. Figure 13-2 shows an example of face images from the AT&T data set. AT&T data set is publicly available at the URL http://www.cam-orl.co.uk/facedatabase.html.

Figure 13-2. Examples of face images from the AT&T data set.

The Yale data set is made up of 11 images per 15 classes/identities (165 total images). Each face is characterized by different facial expressions or configurations: center-light, with/without glasses, happy, left-light, w/no glasses, normal, right-light, sad, sleepy, surprised, and wink. The data set was subdivided into a training set, made up of 5 images per class (75 images), and a test set, made up of 6 images per class (90 images). We repeated our experiments for ten random partitions of the data set and reported the average performances. Figure 13-3 shows an example of face images from the Yale data set. Yale data set is publicly available at the URL: http://cvc.yale.edu/projects/yalefaces/yalefaces.html.

Figure 13-3. Examples of face images from the Yale data set.

In both data sets the face images did not need pre-processing phases, such as re-scaling, rotation or normalization.

4.2 Results with the AT&T Data Set

Table 13-1 reports the results on the AT&T test set.

Table 13-1. Percentage accuracy values on the AT&T test set.

Individual Algorithms		Combined Algorithms			
PCA	LDA	A-KNN	M-KNN	A-NM	M-NM
94.7%	96.1%	95.9%	97.3%	93.3%	96.1%

The average number of principal components for the PCA representation was 119, while we used all 39 components for the LDA representation.

It is worth noting that the best combination result (97.3%) is comparable with those reported in [16] and [17]. In [16], a 97.5% percentage accuracy is reported, but it is averaged only on five runs; in [17], a 99.5% percentage accuracy is reported, but with a rejection rate of 0.5%. Figure 13-2 shows the so called "rank" accuracy, i.e., the percentage accuracy that can be achieved by considering the M identities of the database nearest to the given input face. The input face is considered as correctly recognized if the right identity is one of the M identities. The rank is a reliability measure, and it is very important for video-surveillance applications in uncontrolled environments. Even in this case, the combination of PCA and LDA gives a sharp improvement of the performance, and a better identification reliability and robustness.

Another motivation for fusing PCA and LDA is the average correlation coefficient that we computed between the d^{PCA} and the d^{LDA} vectors. A very low value was obtained: 0.39. This suggests a strong complementarity of the information extracted by the PCA and LDA representations. This confirms that these two approaches are not so correlated as one could think. We think that PCA and LDA are weakly correlated thanks to the good quality of the images in terms of pose and lighting conditions.

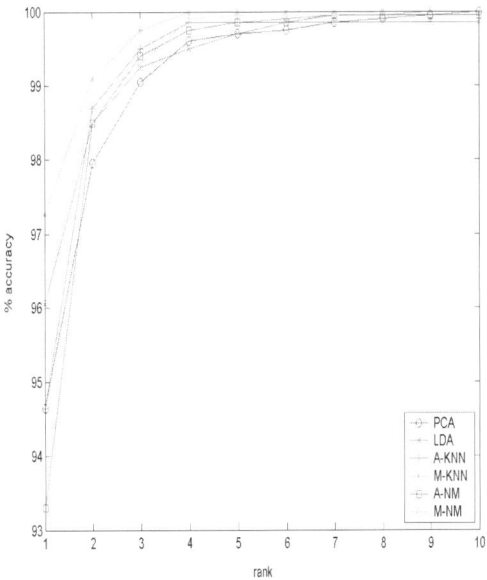

Figure 13-4. Rank curves on the AT&T data set. Reported results show that the combination of PCA and LDA improves the reliability of the system

4.3 Results with the Yale Data Set

While the AT&T data set is characterized by small variations of pose and lighting, the Yale data set is characterized by strong variations of expression and lighting. This task is therefore more complex and the results are obviously worse, even if the number of identities is inferior.

Table 13-2 shows the percentage accuracy of our approaches on this data set.

Table 13-2. Percentage accuracy on the Yale test set.

Individual Algorithms		Combined Algorithms			
PCA	LDA	A-KNN	M-KNN	A-NM	M-NM
83.0%	82.8%	**84.2%**	83.6%	83.6%	81.2%

Even in this case, the combination of PCA and LDA gives the best result. The gain is the same as for the AT&T data set (about 1.3%), but the final result is affected by the performances of PCA and LDA for this difficult task.

The average number of principal components is 33, while we used all 14 components for the LDA representation.

Unfortunately, in this case, we could not compare our results with others, because no work reported in the literature used the Yale data set for combining multiple algorithms for face recognition.

Even in this case, the rank-curves reported in Figure 13-3 show the effectiveness of the decision combination for improving the reliability of a face recognition system.

It should be noted that the average correlation coefficient in this case is high: a value of 0.69 was obtained. In our opinion, PCA and LDA are correlated because of the lighting and face expression variations in the images. The above variations can be considered as "noisy" information that limits the goodness of the feature extraction performed by PCA and LDA. However, the fusion algorithms overcomes partially this limitation. Performance accuracy is superior than that of the best individual recognition algorithm.

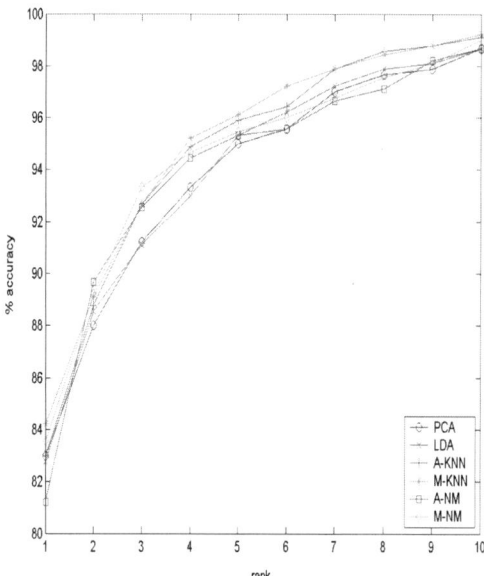

Figure 13-5. Rank curves on the Yale data set. Reported results show that fusion improves the performance of the individual algorithms.

5. CONCLUSIONS

Although many face recognition systems have been proposed in the last years, none of them can completely overcome the limits due to the large variations of critical parameters, such as pose, lighting, scale. In the case of video-based surveillance systems, the poor quality of the acquired images and the variability of the scenarios are other critical problems to be addressed.

The orientation of the research community in this field is to perform recognition when certain constraints are satisfied in terms of pose, lighting, and scenario (still-to-still paradigm), but it is often impossible to guarantee these conditions for real video-surveillance tasks. On the other hand, the approaches based on multiple-stills-to-still and video-to-video paradigms are still a matter of on-going research. At present, an individual face recognition system can achieve good performance only at the expense of robustness and reliability.

In order to improve performances and robustness of face recognition systems for video-surveillance applications, the combination of multiple recognizers was recently proposed. But very few works investigated such fusion.

In this chapter, the fusion of two statistical approaches, namely PCA and LDA, for face representation and recognition has been investigated according to the still-to-still paradigm. Reported results confirm the benefits of such fusion. In particular, for the AT&T data set, these two representations proved to be complementary as shown by the low correlation coefficient. We combined PCA and LDA with the KNN-based combination rule and the NM-based combination rule. In general, the performance of the KNN rule is much better than that of the NM rule: this should mean that the average template (that can be viewed as a low-pass filtering in the domain of the PCA and LDA spaces) reduces the available information. The rank-curves show that the reliability of the recognition always increases with respect to the best individual approach.

Reported results are strongly dependent on the data set. A difficult task like the one presented by the Yale data set shows that the results of the individual classifiers decrease dramatically. However, they can be increased using fusion.

On the basis of the reported results, it is worth devoting further theoretical and experimental investigations to understand the behavior of PCA and LDA, in order to fuse them and to extend their application to real video-surveillance environments.

REFERENCES

[1] A. Jain, R. Bolle, S. Pankanti Eds, *BIOMETRIC – Personal Identification in Networked Society*, Kluwer Academic Publishers, Boston/Dordrecht/London (1999).

[2] H. Wechsler, J.P. Phillips, V. Bruce, F. Folgeman Soulie, T.S. Huang Eds., *Face Recognition – From theory to applications*, Springer, ASI NATO Series, vol.163, 1997.

[3] M.H. Yang, D. Kriegman, N. Ahuja, Detecting Face Images: a Survey, IEEE Trans. on PAMI, 24 (1) 24-58, 2002.

[4] W.Y. Zhao, R. Chellappa, A. Rosenfeld, and P.J. Philips, Face Recognition: a literature survey, *UMD CfAR Technical Report CAR-TR-948*, 2000.

[5] M. Turk, and A. Pentland, Eigenfaces for Face Recognition, *Journal of Cognitive Neuroscience*, 3 (1) 71-86, 1991.

[6] P.N. Belhumeur, J.P. Hespanha, and D.J. Kriegman, Eigenfaces vs. Fisherfaces: Recognition Using Class Specific Linear Projection, *IEEE Trans. on PAMI*, 19 (7) 711-720, 1997.

[7] P.S. Penev and J. Atick, Local Feature Analysis: a general statistical theory for object representation, *Network: Computation in Neural Systems*, 7 (3) 477-500, 1996.

[8] L. Wiskott, J.M. Fellous, N. Krüger, and C. von der Malsburg, Face Recognition by Elastic Bunch Graph Matching, *IEEE Trans. on PAMI*, 19 (7) 775-779, 1997.

[9] M. Lades, J.C. Vorbrüggen, J. Buhmann, J. Lange, C. von der Malsburg, and W. Konen, Distortion invariant object recognition in the dynamic link architectures, *IEEE Trans. on Computers*, 42 (3) 300-311, 1993.

[10] V. Krüger and S. Zhou, Exemplar-based Face Recognition from Video, Proc. of the Fifth *IEEE International Conference on Automatic Face and Gesture Recognition (FGR'02)*, Washington D.C., U.S.A., 2002.

[11] S. Zhou, V. Krüger, and R. Chellappa, Face Recognition from Video: a condensation approach, *Proc. of the Fifth IEEE International Conference on Automatic Face and Gesture Recognition (FGR'02)*, Washington D.C., U.S.A., 2002.

[12] A.J. Howell and H. Buxton, Towards Uncostrained Face Recognition from Image Sequences, *Proc. of the IEEE International Conference on Automatic Face and Gesture Recognition (FGR'96)*, Killington, VT, pp.224-229, 1996.

[13] Y. Li, S. Gong, H. Liddell, Support Vector Regression and Classification Based Multi-view Face Detection and Recognition, *Proc. of the IEEE International Conference on Automatic Face and Gesture Recognition (FGR'00)*, Grenoble, France, pp.300-305, 2000.

[14] F. Roli and J. Kittler Eds., *Multiple Classifier Systems*, Springer Verlag, LNCS 2364, 2002.

[15] B. Achermann and H. Bunke, Combination of Classifiers on the Decision Level for Face Recognition, *Technical Report IAM-96-002*, Institut für Informatik und angewandte Mathematik, Universität Bern, January 1996.

[16] SM. Lucas, Continuous n-Tuple Classifier and its Application to Real-time Face Recognition, *IEE Proceedings of Visual Image and Signal Processing*, 145 (5) 343-348, 1998.

[17] AS. Tolba and AN. Abu-Rezq, Combined Classifier for Invariant Face Recognition, *Pattern Analysis and Applications*, 3 (4) 289-302, 2000.

[18] G.L. Marcialis and F. Roli, Fusion of LDA and PCA for Face Recognition, Proceedings of the Workshop on Machine Vision and Perception, 8th *Workshop of the Italian Association for Artificial Intelligence (AIIA'02)*, available at the URL: http://www-dii.ing.unisi.it/aiia2002.

[19] G.L. Marcialis and F. Roli, Fusion of LDA and PCA for Face Verification, *Proceedings of the Workshop on Biometric Authentication*, M. Tistarelli, J. Bigun and A.K. Jain Eds., Springer LNCS 2359, Copenhagen, Denmark, pp.30-37, 2002.

[20] R.O. Duda, P.E. Hart, and D.G. Stork, *Pattern Classification*, John Wiley & Sons, USA 2001.

[21] W. Zhao, A. Krishnaswamy, R. Chellappa, D. Swets, and J. Weng, Discriminant Analysis of Principal Components for Face Recognition, *in Face Recognition: From Theory to Applications*, Eds. H. Wechsler, P.J. Phillips, V. Bruce, F.F. Soulie and T.S. Huang, Springer-Verlag, pp. 73-85, 1998.

[22] L.F. Chen, H.Y.M. Liao, M.T. Ko, J.C. Lin, G.J. Yu, A new LDA-based face recognitions system which can solve the small sample size problem, *Pattern Recognition*, 33 1713-1726, 2000.

[23] F. Goudail, E.Lange, T. Iwamoto, K. Kyuma, N. Otsu, Face recognition system using local autocorrelation and multiscale integration, *IEEE Trans. PAMI*, 18 (10) 1024-1028, 1996.

[24] K. Fukunaga, *Introduction of Statistical Pattern Recognition*, Academic Press, New York, 1990.

[25] K. Liu, Y.Q. Cheng, J.Y. Yang, X. Liu, A generalised optimal set of discriminant vectors, *Pattern Recognition*, 25 (7) 731-329, 1992.

[26] J. Yang, J. Yang, Why can LDA be performed in PCA transformed space?, *Pattern Recognition*, 36 (2) 563-566, 2003.

Chapter 14

INFORMATION THEORY BASED FACE TRACKING

Evangelos Loutas, Christophoros Nikou and Ioannis Pitas
University of Thessaloniki, Department of Informatics

Keywords: Mutual information, arbitration scheme, occlusion, illumination changes.

1. INTRODUCTION

Automatic detection and tracking of human body parts (e.g. face, arms) is a challenging research topic with applications in many domains such as human computer interaction, surveillance, face recognition and human joint audio and video localization systems.

In this framework, Bayesian approaches express the posterior probability of the motion parameters in terms of a prior probability and a likelihood function [1]. The prior probability is representative of the previous history of the tracked object and the likelihood express its similarity to an appearance based model learnt through statistical training and is representative of the present status of the tracked object.

The main characteristics of the relevant published work are the use of an image model learnt through statistical training and the fusion of different tracking cues. An appearance model consisting of a stable component, a transient component and an outlier detection process is proposed in [2]. Object tracking is performed using color, texture, and edge information in [3], while edge and ridge information is used in [4]. Grayscale information and motion models are combined in [5] to perform tracking of 3D articulated figures.

Head orientation is calculated by using either feature based methods [6][7] or appearance based methods [8][9]. The latter rely on using training sets of face images under various poses, while the former do not require statistical training. Appearance based methods are particularly interesting, as they can be combined in a probabilistic framework to obtain a single perceptual output.

The face tracking scheme proposed in this paper relies on calculating the probability of motion parameters as the product of a facial observation probability obtained using an appearance based model, a mutual information tracking cue and a temporal model. The novelty of our approach lies in the use of mutual information as a separate cue in a probabilistic face tracking framework. Furthermore, the probability of face observation is constrained using a temporal model based on the automatically generated feature point sets.

Head orientation calculation is performed using a mutual information based scheme as well. The proposed approach does not require training for head orientation estimation and has produced good results in determining pose under facial appearance changes and illumination variations.

The tracking algorithm is initialized using a probabilistic framework and is interpreted as a probabilistic face detector. An arbitration scheme is also used to obtain an extension of the algorithm to cover multiple face cases.

The main contributions of the current work are the use of a novel probabilistic model based on automatically generated feature point sets in an object tracking scheme, the introduction of mutual information as a separate cue in a probabilistic object tracking and the head orientation calculation using mutual information.

The proposed tracking scheme was tested on real image sequences. The tracker performs well in partial occlusion and illumination changes, because it combines the robustness of the mutual information systems to illumination changes and the appearance based face detection systems to partial occlusion.

The remainder of the chapter is organized as follows: The probability of facial observation estimation and the tracking initialization procedure are described in section 2. The mutual information tracking cue is presented in section 3. The temporal model construction is presented in section 4. The tracking process is described in section 5. Experimental results are presented in section 6 and conclusions are drawn in section 7.

2. PROBABILITY OF FACIAL OBSERVATION ESTIMATION

The acquisition of the probability of facial observation estimates is an important part of a probabilistic face tracking framework. The probability model of the facial observations is learnt through training using automatically generated feature point sets. Each image of the training set is reduced to a set of automatically generated feature points [10][11]. The feature points represent image corners and are characterized by large gradient variations in both horizontal and vertical directions.

2.1 Face feature generation and training

The feature points $\mathbf{v}_f = [x, y]^T$ [10], are generated using a matrix:

$$\mathbf{Z} = \begin{bmatrix} \sum_w J_x^2 & \sum_w J_x J_y \\ \sum_w J_x J_y & \sum_w J_y^2 \end{bmatrix} \quad (14.1)$$

where J_x and J_y are the image gradients along the x and y direction respectively and W is a $n \times n$ window centered on the candidate feature point. Matrix $\mathbf{Z}$ is by definition a zero-positive one and is calculated for every candidate feature point. with two eigenvalues $\lambda_1 > \lambda_2 \geq 0$. Feature points having two large eigenvalues of their matrix $\mathbf{Z}$ are selected. Furthermore, the geometrical distance between two feature points must not be smaller than a predefined threshold (the so-called feature point neighborhood threshold) to ensure that the feature points do not concentrate on small strongly textured image neighborhoods. The feature point set contains N feature points. Most of them represent corners generated by the intersection of the object contours or corners of the local intensity pattern not corresponding to obvious scene features [12]. In the case of faces, the feature point set is expected to lie on face areas containing intensity variations such as the face contour, the eyes area, the nose area and the mouth area (see Figure 14-1).

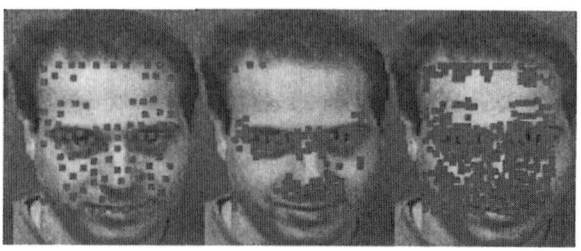

Figure 14-1. (a) Feature point sets of 100 feature points neighborhood threshold=5. (b) Feature point set of 100 feature points, neighborhood threshold=3. (c) Feature point set of 300 feature points. Neighborhood threshold=3.

The training procedure involves the feature point set generation from a number of training images. The "ORL Database of Faces"[13] containing a total number of 400 images of 40 different persons was used for training. Feature point sets were generated on the facial region of each training image. The facial region bounding rectangle was manually defined during the training process. The feature point set generation process was performed inside the manually defined facial region for each image belonging to the training set.

As stated in [14], a major difficulty in the application of statistical feature point training is the efficient establishment of a rough registration between the different instances of the training set.

Therefore, to avoid tedious manual interaction on very large feature point sets we have resorted to a semi-automatic registration procedure. A bounding box containing the face was drawn on each image of the training set and registered with respect to the bounding box of an arbitrarily chosen reference image. Fully automatic unsupervised registration algorithms also proposed in literature, may be applied but their output strongly depends on the similarity metric used [15].

Let $\mathbf{v}_{r_i}^1$ be the i-th feature point geometrical coordinates with respect to the upper left corner of the bounding box, belonging to the first image of the training set and $\mathbf{v}_{r_i}^l$ the geometrical coordinates with respect to the upper left corner of the bounding box of a feature point belonging to the l-th image of the training set and has not been matched yet. We have assumed correspondence for features $\mathbf{v}_{r_i}^1$ and $\mathbf{v}_{r_j}^l$ with j^* satisfying:

$$j^* = \arg\min_j \left\| \mathbf{v}_{r_i}^1 - \mathbf{v}_{r_j}^l \right\| \qquad (14.2)$$

among the feature points of image l not yet matched.

14. Information theory based face tracking

The feature point generation can be seen as a mapping procedure. Each image of the training set is mapped to a "feature point set" space with reduced dimensionality. The number of feature points N is selected to be much less than the total number of image pixels N_1, $N \ll N_1$. It is convenient to set $N < N_T$, where N_T is the cardinality of the training image set. In our case $N_T = 400$. A second dimensionality reduction step can be accomplished by using standard PCA. This step is necessary if further dimensionality reduction is desired without reducing the number of feature points N.

The Gaussian probability density function of facial observation can be computed using the first M principal components, typically $M = 0.2N$. If N is chosen as $N \cong 0.1N_1$ then $M \cong 0.02N_1$. This corresponds to a significant data reduction of a factor of 50:1. The level of reduction can be controlled by appropriately selecting the number of feature points N. Let $J(\mathbf{v}_{f_1})$ be the image intensity at pixel $\mathbf{v}_{f_1}$. The feature vector $\mathbf{x} = [J(\mathbf{v}_{f_1}),\ldots,J(\mathbf{v}_{f_N})]^T$ can be expressed as:

$$\mathbf{x} = \overline{\mathbf{x}} + \mathbf{Pb}, \tag{14.3}$$

where $\mathbf{P}$ is the matrix whose columns are the eigenvectors of the covariance matrix:

$$\mathbf{C} = E[(\mathbf{x} - \overline{\mathbf{x}})(\mathbf{x} - \overline{\mathbf{x}})^T], \tag{14.4}$$

$\overline{\mathbf{x}}$ is the mean feature vector:

$$\overline{\mathbf{x}} = \frac{1}{N_T}\sum_{i=1}^{N_T}\mathbf{x}_i \tag{14.5}$$

and

$$\mathbf{b} = \mathbf{P}_M^T (\mathbf{x} - \overline{\mathbf{x}}) \tag{14.6}$$

are the coordinates of $\mathbf{x} - \overline{\mathbf{x}}$ in the eigenvector basis. A principal component feature vector $\mathbf{y} = [y_1,\ldots y_M]^T$ is obtained by:

$$\mathbf{y} = \mathbf{P}_M^T (\mathbf{x} - \overline{\mathbf{x}}) \tag{14.7}$$

where $\mathbf{P_M}$ is a submatrix of $\mathbf{P}$ containing the M principal eigenvectors.

2.2 Probability of facial observation estimation

The estimation of the facial observation probability, using the multiscale extension of the face detection procedure presented in [16], is accomplished by calculating the likelihood $p(\mathbf{x}|\underline{\phi}_t,\Omega)$ of a target, where $\mathbf{x}$ is the input pattern in the "feature point set space" representing the data observed, $\underline{\phi}_t$ is a vector containing face location, rotation (orientation) and scaling parameters and Ω represents the face class. Using the results obtained by PCA, $p(\mathbf{x}|\underline{\phi}_t,\Omega)$ can be approximated with [16]:

$$\hat{p}(x|\underline{\phi}_t,\Omega) = p_M(x|\underline{\phi}_t,\Omega)\hat{p}_{N-M}(x|\underline{\phi}_t,\Omega) \qquad (14.8)$$

where $p_M(\mathbf{x}|\underline{\phi}_t,\Omega)$ is the term estimated from the M principal components and $\hat{p}_{N-M}(\mathbf{x}|\underline{\phi}_t,\Omega)$ is the estimated contribution of the remaining components.

In order to estimate $\hat{p}(\mathbf{x}|\underline{\phi}_t,\Omega)$ over a new image region, a set of feature points should be generated using the previously described algorithm. An estimate of the face position and scale is thus obtained. The probability $\hat{p}(\mathbf{x}|\underline{\phi}_t,\Omega)$ of a pattern $\mathbf{x}$ belonging to a face is generally normalized with respect to its maximum value $\hat{p}(\mathbf{x}|\underline{\phi}_t,\Omega)_{max}$. The normalized probability is compared to a predefined threshold in order to perform facial region labeling.

2.3 Tracking algorithm initialization

The face tracking algorithm initialization procedure is based on the estimation of the facial observation probability, which is extended to handle multiple faces. Candidate facial regions are considered all regions, whose normalized face observation probability exceeds a predefined threshold. In order to eliminate false facial region candidates an arbitration scheme similar to that presented in [17] is implemented. The steps of the initialization of the multiple face tracking algorithm are:
1. Calculate the facial observation probabilities over the entire image (eq. 14-8).
2. Reject the candidate regions whose normalized facial observation probability is below a predefined threshold. Mark these candidate regions as non face regions.

- **Repeat**
  i. Mark as a face the unmarked image region assigned to the maximum facial observation probability.
  ii. Perform *the arbitration scheme*:
   1. Reject any candidate facial region whose center lies within a previously defined facial region.
   2. Reject any candidate facial region overlapping with a previously defined facial region.
   3. Reject any candidate facial region when the number of less probable candidate facial regions within them is less than a predefined threshold.
- **Until** all candidate regions are marked as face or non face.

3. MUTUAL INFORMATION CUE

The tracking process can be modeled as a communication between a transmitter (the reference face region A_{t-1} at time instant $t-1$) and a receiver (the target face region A_t at time instant t) with a N_{max} symbol alphabet (the maximum number of grayscale levels). Mutual information is a measure of the amount of information transmitted through the communication channel. Let $U(\underline{\phi}_{t-1}), V(\underline{\phi}_t)$ be two random variables of the images having marginal probability mass functions $p(u)$, $p(v)$ and $u_i = J_1(\mathbf{p}_{f_{t-1}}), v_j = J_2(\mathbf{p}_{f_t})$, J_1 and J_2 are the reference and target images respectively, $\mathbf{p}_{f_{t-1}} \in A_1, \mathbf{p}_{f_t} \in A_2$ and $\underline{\phi}_t = [\mathbf{v}_{tot_t}, s_t, \vartheta_t]^T$ is the tracked face parameter vector at time t. $\mathbf{v}_{tot_t}$, contains the geometrical coordinates of the feature point set at time t, $\mathbf{v}_{tot_t} = [\mathbf{v}_{f_1}, \mathbf{v}_{f_2}, \ldots, \mathbf{v}_{f_N}]^T$, while s and ϑ represent scale and rotation parameters respectively.

The mutual information of two random variables U, V with a joint probability mass function $p(u,v)$ is defined as [18]:

$$I(U(\underline{\phi}_{t-1}), V(\underline{\phi}_t)) = \sum_{i=1}^{N_{max}} \sum_{j=1}^{N_{max}} p(u_i v_j) \log_2 \frac{p(u_i v_j)}{p(u_i)p(v_j)} \quad (14.9)$$

The maximum mutual information for a particular prior $p(u)$ is [19]:

$$I_{max}(U(\underline{\phi}_{t-1}),V(\underline{\phi}_t)) = -\sum_{i=1}^{N_{max}} p(u_i) log_2 p(u_i) \qquad (14.10)$$

and reaches its maximum value when:

$$p(u_i) = \frac{1}{log_2 N_{max}}, 0 \leq i < N \qquad (14.11)$$

Let the probability based on the mutual information tracking cue be:

$$p_{MI}(U,V|\underline{\phi}_t,\underline{\phi}_{t-1}) \stackrel{\Delta}{=} \frac{I(U(\underline{\phi}_{t-1}),V(\underline{\phi}_t))}{I_{max}(U(\underline{\phi}_{t-1})V(\underline{\phi}_t))} \qquad (14.12)$$

Since $I(U,V) \geq 0$ [18], $0 \leq p_{MI}(U,V|\underline{\phi}_t,\underline{\phi}_{t-1}) \leq 1$. A large value of $p_{MI}(U,V|\underline{\phi}_t,\underline{\phi}_{t-1})$ indicates a strong match between the reference and the target regions, while a small value of $p_{MI}(U,V|\underline{\phi}_t,\underline{\phi}_{t-1})$ indicates a weaker match.

4. TEMPORAL MODEL

The temporal model describes the probability of face appearance given its location at the previous time instance. The temporal model is used as a location constraint factor [5] in the tracking process. Scale variation s is modeled as a Gaussian distribution:

$$p(s_t | s_{t-1}) \approx c_1 e^{-c_2(s_t - s_{t-1})^2} \qquad (14.13)$$

In order to model the facial position variation, the feature point sets generated on the reference and target regions are used. The overall facial position variation is modeled by using the feature point geometrical coordinates as:

14. Information theory based face tracking

$$p(\mathbf{v}_{tot\,t} | \mathbf{v}_{tot\,t-1}) \approx c_3 e^{-c_4 \sum_k (v_x^k(t)-v_x^k(t-1))^2 + (v_y^k(t)-v_y^k(t-1))^2} \quad (14.14)$$

where $v_x^k(t)$ $v_y^k(t)$ are the x and y coordinates of feature point k at time instant t respectively.

Finally, rotation is modeled by:

$$p(\vartheta_t | \vartheta_{t-1}) \approx c_5 e^{-c_6(\vartheta_t - \vartheta_{t-1})^2} \quad (14.15)$$

Constants c_1, c_2, c_3, c_4, c_5 and c_6 are empirically determined. Too small values of c_2, c_4 and c_6 will reduce the importance of the temporal model for the tracking process [1]. The overall temporal model term is obtained as the product of the terms $p(s_t | s_{t-1})$, $p(\mathbf{v}_{tot\,t} | \mathbf{v}_{tot\,t-1})$, $p(\vartheta_t | \vartheta_{t-1})$.

$$p_{TEMP}(\underline{\phi}_t | \underline{\phi}_{t-1}) = p(s_t | s_{t-1}) \, p(\mathbf{v}_{tot\,t} | \mathbf{v}_{tot\,t-1}) \, p(\vartheta_t | \vartheta_{t-1}) \quad (14.16)$$

5. FACE TRACKING

The detected faces are tracked by maximizing the probability obtained as the product of the probability of facial observation, the mutual information tracking cue probability and the temporal model probability term:

$$p_{TRACKING} = c_7 p_{MI}(U,V | \underline{\phi}_t, \underline{\phi}_{t-1}) p_{TEMP}(\underline{\phi}_t | \underline{\phi}_{t-1}) p(\mathbf{x} | \underline{\phi}_t, \Omega) \quad (14.17)$$

The term c_7 is a normalizing factor, while the term $p_{MI}(U,V | \underline{\phi}_t, \underline{\phi}_{t-1})$ represents the mutual information tracking cue probability and $p_{TEMP}(\underline{\phi}_t | \underline{\phi}_{t-1})$ represents the temporal model term. The probability $p_{TRACKING}$ has a resemblance to the Bayesian formulation, without being characterized as Bayesian in strict mathematical sense.

In order to obtain the final estimate of the head orientation, we assume at first only translation and no rotation or scaling. The estimate is then refined by estimating the scale factor and the rotation angle. Better results may be obtained by adopting a recursive refining process.

6. EXPERIMENTAL RESULTS

The proposed algorithm has been tested on a variety of real face image sequences under different illumination and occlusion conditions. Results on a single face sequence without illumination changes or occlusion are presented in Figure 14-2.

Figure 14-2. Tracking results under constant illumination conditions.

As can be observed, the face position and orientation are correctly determined. Tracking results on a similar sequence with illumination changes are presented in Figure 14-3.

Figure 14-3. Tracking results under varying illumination conditions

A slight drift in the estimated facial position is noticed in very dark image sequences when the tracking process is prolonged for too long. Results on multiple face image sequences suffering from lightening changes and partial occlusion are presented in Figures 14-4 and 14-5 respectively.

14. Information theory based face tracking

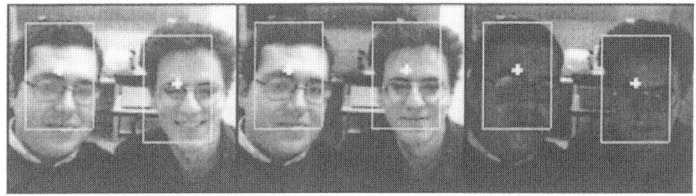

Figure 14-4. Tracking results in a two face image sequence under varying illumination conditions

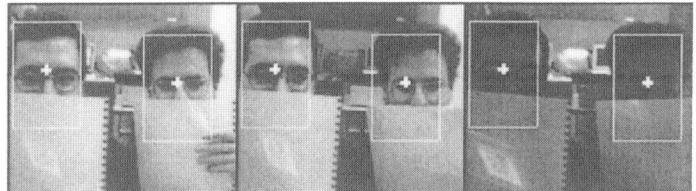

Figure 14-5. Tracking results in a two face image sequence having illumination variations and partial occlusion

The facial positions are correctly determined in the multiple face case even under severe partial occlusion and illumination changes. In general, the face tracking algorithm proposed in this chapter can effectively track multiple faces under significant illumination changes and partial occlusion.

7. CONCLUSIONS

A mutual information based face tracking scheme was presented in this chapter. The facial observation probability estimation is performed using sets of automatically generated feature points. A mutual information tracking cue and a temporal model term are also used.

The main contributions of the proposed scheme are the introduction of a novel appearance based model and the use of a mutual information tracking cue which is combined with a temporal model. Moreover, the implementation of an arbitration scheme, to face tracking initialization is also important since it allows a multiple face tracking extension.

The proposed algorithm was tested on real face sequences. Results have shown that the facial position is correctly determined even in image sequences having rather strong illumination changes and partial occlusion. The face orientation was correctly determined under normal illumination conditions and slight illumination changes. Robustness to illumination changes is obtained by using the mutual information tracking cue, while robustness to partial occlusion is obtained by the use of the appearance based model.

ACKNOWLEDGEMENTS

This study has been partially supported by the Commission of the European Communities, in the framework of the project IST-1999 20993 CARROUSO (Creating, Assessing and Rendering of High Quality Audio-Visual Environments in MPEG-4 context).

REFERENCES

[1] J. Ruanaidh and W. Fitzgerald, *Numerical Bayesian methods applied to signal processing*, Springer-Verlag, 1996.

[2] A. Jepson, D. Fleet, and T. Maraghi," Robust online appearance models for visual tracking" in *Proc. of 2001 Int. Conf. On Computer Vision and Pattern Recognition*, 2001, vol I, pp. 415-422.

[3] C. Rasmussen and G. D. Hager, "Probabilistic data association methods for tracking complex visual objects", *IEEE Transactions on Pattern Analysis and Machine Intelligence*, vol. 23, no 6,pp. 560-576, 2001.

[4] H. Sidenbladh and M. Black, "Learning image statistics for Bayesian tracking", in *IEEE International Conference on Computr Vision (ICCV)*, Vancouver, Canada., 2001, vol. 2, pp. 709-716.

[5] H. Sidenbladh, F. De la Torre, and M. Black, "A framework for modeling the appearance of 3d articulated figures", *in IEEE International Conference on Automatic Face and Gesture Recognition (FG)*, Grenoble, France., 2000, pp. 368-375.

[6] T. Jebara and A. Pentland, "Parametrized structure from motion for 3d adaptive feedback tracking of faces", *in Proceedings of International Conference on Computer Vision and Pattern Recognition*, 1997, pp. 144-150.

[7] A. Nikolaidis and I. Pitas, "Facial feature extraction and pose determination", *Pattern Recognition*, vol. 33, no. 11, pp. 1783-1791, 2000.

[8] T. Darrell, b. Moghaddam, and A. Pentland, "Active face tracking and pose estimation in an interactive room", in *Proceedings of the International Conference on Computer Vision and Pattern Recognition*, 1996, pp. 67-72.

[9] Y. Wu and K. Toyama, "Wide-range, person and illumination-insensitive head orientation estimation", *in IEEE International Conference on Automatic Face and Gesture Recognition (FG)*, Grenoble, France., 2000, pp. 183-188.

[10] C Tomasi and T. Kanade, *Shape and Motion from Image Streams: a Factorization Method – Part 3 Detection and Tracking of Point Features*, 1991.

[11] K. Rohr, *Landmark-based image analysis*, Kluver Academic Publishers, 2001.

[12] A. Verri E. Trucco, *Introductory techniques for 3-D Computer Vision*, Prentice Hall, 1998.

[13] F. Samaria and A. Harter, "Parameterisation of a stochastic model for human face identification", in *Proceedings of 2nd IEEE Workshop on Applications of Computer Vision*, Sarasota FL, 1994, pp. 138-142.

[14] T. F. Cootes and C. J. Taylor, "Active shape models – their training and application", *Computer Vision and Image Understanding*, 1(1): 38-59, 1995

[15] F. Heitz C. Nikou and J. P. Armspach "Robust registration of dissimilar single and multimodal images", vol. 2, pp. 51-65, 1998.

[16] B. Moghaddam and A. Pentland, "Probabilistic visual learning for object representation", *IEEE Transactions on Pattern Analysis and Machine Intelligence*, vol. 19, no. 7, pp. 696-710, 2001.

[17] H. Rowley, S. Baluja, and T. Kanade, "Neural network-based face detection", *IEEE Transactions on Pattern Analysis and Machine Intelligence*, vol. 20, no. 1, pp. 23-37, 1998.

[18] S. Haykin, *Communication Systems*, J. Wiley, 1994.

[19] M. Skouson, Q. Guo and Z. Liang, "A bound on mutual information for image registration", *IEEE Transactions on Medical Imaging*, vol. 20, no 8, pp. 843-846, 2001

Chapter 15

OPTIMUM FUSION RULES FOR MULTIMODAL BIOMETRIC SYSTEMS

Lisa Osadciw, Pramod Varshney, Kalyan Veeramachaneni
Syracuse University, Syracuse, USA

Keywords: Biometrics, data fusion, fingerprint, iris scanning, hand geometry

1. INTRODUCTION

"Biometrics is the automated use of physiological or behavioral characteristics to determine or verify identity". The first modern commercial biometric device was introduced over 25 years ago when a machine that measured finger length was installed for maintaining employee time records at Shearson Hamil on Wall Street. In the ensuing years, hundreds of these hand geometry devices have been installed for security purposes at facilities operated by Western Electric and Naval Intelligence. In the US today, biometric security systems may be found in the Oakland International Airport (face recognition), Chicago O'Hare International (fingerprint), and Navy Consolidated Building (Iris recognition) [1]. With the demand for better security technology increasing, a variety of pilot projects have been completed recently in the area of access control [2]. These projects highlight the need to improve biometric verification accuracy.

This chapter will describe issues preventing widespread deployment of biometric security systems and then present an approach that is based on multisensor fusion to mitigate the existing problems. A focus is placed on deriving the optimum Bayesian fusion rule based on the system requirements and the impact various types of errors have on system security. Other researchers have published numerous experimental results demonstrating the effectiveness of fusion in biometrics [2-7]. These studies focus on the promising results rather than rule formation. This chapter presents a comprehensive approach to designing a Bayesian fusion rule for multimodal biometric features. The goal is to improve the accuracy of biometric verification.

2. WHY BIOMETRICS?

Personal safety in public and private buildings has always been a concern but since September 11, 2001 is receiving more attention. Identification methods can be grouped into three classes: something you possess as in an ID card, something you know, and something unique about you like biometrics. Possessions (e.g. keys) can be easily lost, forged or duplicated. Knowledge can be forgotten as well as shared, stolen, or guessed. The cost of forgotten passwords is high and accounts for 40% - 80% of all the IT help desk calls [8]. Resetting the forgotten or compromised passwords costs as much as 340$/user/year [9]. Biometrics, on the other hand, are inherently secure since they are some unique feature the person physically has. The science of biometrics is an elegant solution to identifying an individual and avoids problems faced by possession-based and knowledge-based security approaches.

The aggregate security level of a system increases as these three identification approaches are combined in various ways. The least secure approach is based on PINs (Personal Identification Numbers), which can be easily guessed. The system's security level can be improved by adding some possession such as an identification card. An identification card with a single biometric improves security further. Finally, an identification card with multiple biometrics supports a very high security level as illustrated in Figure 15-1.

15. Optimum fusion rules for multimodal biometric

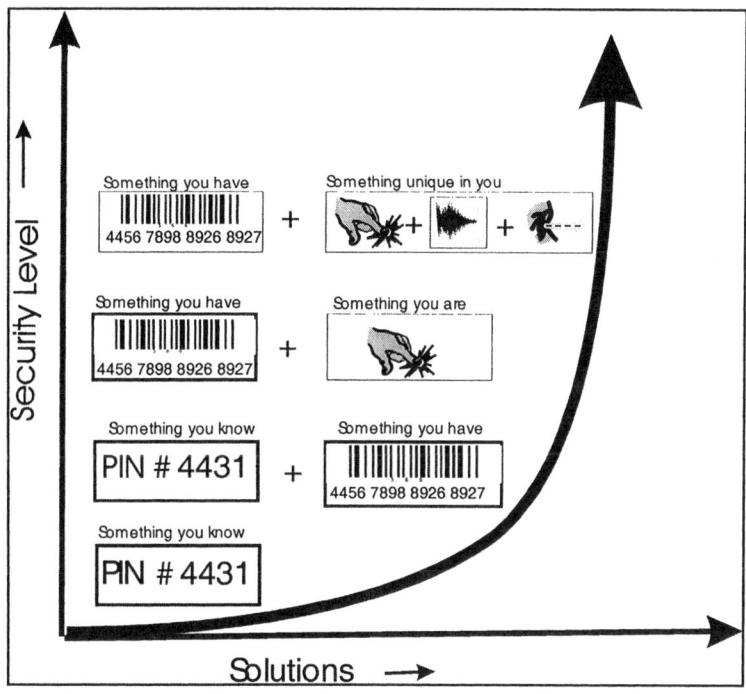

Figure 15-1. Solutions to Increasing Security Needs

3. EXISTING BIOMETRIC TECHNOLOGIES

Currently, the five most common biometric technologies are fingerprinting, iris scanning, hand geometry comparison, face recognition and voice verification. These techniques have significantly varying degrees of accuracy, ease of use, failure to enroll, failure to acquire, and universality. Each technology must perform four basic tasks: biometric acquisition, feature extraction, matching, and decision making.

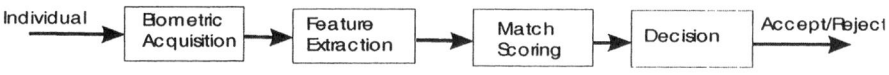

Figure 15-2. Biometric Identification Process

3.1 Fingerprinting

Fingerprinting technology has been in the market for a while now and has become more convenient with the emergence of several inkless scanning technologies. The exponential increase in processor performance has greatly reduced processing time as well. The basic principle of fingerprinting is to identify an individual by the unique ridge structure on a finger. There are different methods of acquiring the ridge structure namely: CMOS capacitance, thermal, ultrasound. Different identification techniques focus on various ridge structures in order to identify an individual.

The most commonly used fingerprint features are ridge bifurcations and ridge endings, collectively known as minutiae. These are referred to as biometric features that are extracted from the acquired image. The feature extraction process starts by examining the quality of the input gray-level image. Usually this is accompanied by computations indicating the orientation of the fingerprint image referred to as the orientation field. This field reflects the local ridge direction at every pixel. The orientation field is sometimes used to tune filter parameters for enhancement and ridge segmentation. From the segmented ridges, a thinned image is computed so that the minutiae features may be located. Usually, a minutia post-processing stage is necessary to clean up spurious minutiae resulting from either enhancement, ridge segmentation or thinning artifacts. Then a distance between two fingerprint feature sets is used as a score to accurately and reliably identify the individual. A decision is made using the distance as a score to decide if the person has been identified or not.

Skin conditions and uniqueness of features can prevent this biometric from being useful. Dirt, ink, food, etc. on the fingerprints will prevent a successful match as well as possibly degrading the fingerprint acquisition device for others. Some individuals have no strongly identifiable fingerprint minutiae. Thus although this is a very successful and mature technology, other forms of biometric identification must be developed.

3.2 Iris Scanning

Iris scanning is a relatively new technology. The colored part of the eye bounded by the pupil and sclera is the iris. This part of the eye is extremely rich in texture with potentially 256 features to match. The accuracy of this system is as high and sometimes higher than fingerprinting. The main drawback is that the iris recognition system is not as user friendly as the others. Some feel it is invasive while others have difficulty remaining still enough for the biometric to be acquired. Glasses and con-tacts can significantly inhibit this technology.

3.3 Hand Geometry

Hand geometry based authentication is an extremely user friendly biometric with low accuracy. The lengths of the fingers as well as other hand shape attributes are extracted from images and used in the feature representation of the hand. A relatively inexpensive camera can be used to acquire the features resulting in a low cost of a system. The system is easy to build with easy computations. Hand geometry systems, however, have relatively higher FAR and FRR rates. In a hand recognition system, jewelry such as rings can cause errors. Hand injuries also adversely affect the system accuracy.

3.4 Face Recognition

Human beings have always naturally used face recognition for personal identification purposes. Applications like biometrics, content-based information retrieval, visual surveillance and human computer interaction necessitate successful automation of the recognition task. In automatic face recognition, computer systems are employed to match the test (newly acquired and unknown) face image against a collection of known face images (training faces) in the database. Although the recognition task seems to be easy and straightforward for people, automated face recognition system becomes challenging and difficult. This is primarily due to the inherent variations in the image acquisition process in terms of image quality, geometry, illumination effects, and occlusion (glasses, facial hair, etc.). These major problems currently limit the accuracy of face recognition.

Face recognition is a very attractive as a biometric because the data is already by many of us in the form of a passport or driver's license. Secondly, this biometric can easily be captured by an ordinary camera. Also, the surveillance systems can rely on capturing image without the cooperation of the user. There are many inherent qualities that make it beneficial to automate and improve face recognition.

3.5 Voice Verification

Voice verification, like face recognition, is attractive because of its prevalence in human communication, and its accuracy is currently limited. Speaker identification suffers considerably from variations in the microphone and/or the transmission channel. The performance deteriorates badly as enrollment and use conditions become increasingly mismatched. Background noise can also be a considerable problem. Variations in voice due to illness, emotion or aging are other problems requiring further research.

4. ISSUES WITH EXISTING TECHNOLOGIES

4.1 Failure to Enroll

The failure to enroll rate measures the proportion of individuals for whom the system is unable to generate repeatable samples. This includes those unable to present biometric feature (for example, the iris system may fail to enroll the iris of a blind eye). Also those unable to produce an image of sufficient quality at enrollment, as well as those unable to reproduce their biometric feature consistently are included in this rate. Fingerprint based systems have been found to have 2%, and iris systems a 0.5% failure to enroll rate [10].

4.2 Failure to Acquire

The failure to acquire rate measures the proportion of attempts for which the system is unable to capture or locate an image of sufficient quality. This includes cases where the user is unable to present the required biometric feature (e.g. having a plaster covering his or her fingerprint). The voice biometric also fails when a user catches cold and cannot produce voice of sufficient quality. The voice biometric is the most vulnerable to this failure 2.5%[10], followed by 0.8% for fingerprint.

4.3 Accuracy

Accuracy refers to the rates that two types of errors occur. The biometric system makes 1 of 4 possible decisions during the matching process. The possible decisions are
 1. The genuine employee is accepted.
 2. The genuine employee is rejected.
 3. The imposter is accepted.
 4. The imposter is rejected.

Out of these, there are two types of errors called the false rejection error, number two, and false acceptance error, number three. Performance of the biometric systems is measured by their accuracy in identification, which is calculated using false rejection and false acceptance errors. The first, which is also often used as the only performance measure, is called genuine acceptance rate.

The system designer exercises a tradeoff between false acceptance rate, FAR, and false rejection rate, FRR, since both of them cannot be reduced simultaneously. Varying the operating point or decision threshold of the biometric system makes this tradeoff. Since security is usually the prime objective, a low FAR is usually selected at the expense of a high FRR. This causes a variety of problems such as long transactions times, extensive user training, as well as employees circumventing the system to avoid the daily frustration of using it.

4.4 Universality

Universality is another major problem hindering the implementation of large-scale biometric systems. This problem results from the inability to collect a distinguishing biometric feature from a segment of the population that the system is intended. For example, there are difficulties associated with successfully collecting identifying features from an Iris scan of a blind eye. Also, Asian women frequently have indistinguishable fingerprints. An important characteristic of a biometric technology is its ability to distinguish individuals in an entire population. If the system must be avoided for certain individuals, a security weakness results.

4.5 Transaction Time

Another major factor affecting user acceptance of a biometric security system is the transaction time or time required to collect the biometric data. The transaction times in Table 1 were experimentally determined for various biometric technologies [10]. Users seem to be sensitive to transaction times exceeding 30 seconds. An approach is presented in this chapter that combines multiple modalities. In order to maintain acceptable transaction times, the modalities are chosen so that the system can collect multiple features simultaneously. For example, the face and voice have the longest transaction times but the mean transaction time can be reduced from 37 seconds to 22 seconds if collected together. One drawback is that this may require more training for the users. The user interface is another area currently needing more research in biometrics.

Table 15-1. User Transaction Time

Modality	Mean (s)	Median (s)	Minimum (s)
Face	15	14	10
Hand	10	8	4
Voice	12	11	10
Iris	12	10	4
Fingerprint (Optical)	9	8	2

5. MULTIMODAL BIOMETRICS

Recently there has been a lot of interest in multimodal biometric identification systems [5][10]. As discussed in previous sections, each modality has its own limitations, issues, and problems. Not all of these can be solved for a single biometric through further research. Another approach to building a more robust biometric security system involves integrating multiple sensors. These multimodal biometric systems can be broadly categorized based on the point from the origin of the data collected from the sensor in Figure 15-2. These categories are feature level fusion, score level fusion, and decision level fusion and shown in Figure 15-3.

5.1 Feature Level Fusion

Feature level fusion uses data collected after the feature extraction processing of Figure 15-2. At this level, features extracted using multiple sensors are concatenated. An example is the fusion of fingerprints collected using both an optical and ultrasound sensor. This is data has minimal processing yielding large amounts of data. This is the first level of fusion as shown in Figure 15-3. This fusion level requires the largest amount of communication bandwidth and the most fusion processing. However, fusion can potentially enhance performance if feature level fusion is properly done.

5.2 Score Level Fusion

The second fusion point is after the score processing typically accomplished using signal and/or image processing techniques. The matching scores reported by multiple matchers are combined to support a decision in the fusion processor [7]. Score level fusion enhances system performance while requiring lower communication bandwidths.

15. Optimum fusion rules for multimodal biometric

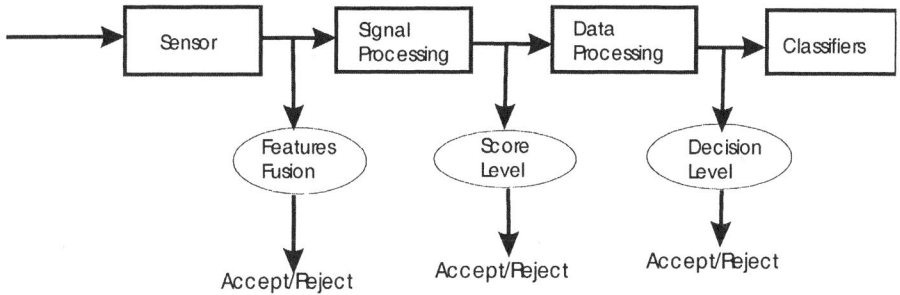

Figure 15-3. Illustration of Fusion Levels

5.3 Decision Level Fusion

Decision level fusion is at the most abstract level, where multiple accept/reject decisions of the multiple systems are consolidated into one decision. This chapter focuses on this fusion level. First, this level takes advantage of the tailored processing performed by each biometric sensor. It also requires the lowest communication bandwidth. Thus, this fusion level supports system scalability.

6. MULTIMODAL FUSION APPROACHES

Some research has successfully fused multimodal biometrics and studied its benefits for personal identification [3,4,10,12]. These fusion approaches focused on obtaining empirical results and not on the formation of the optimal rule from the complete set of fusion rules. However for example, the BioID [10] does successfully choose from different fusion strategies to vary the system security levels. The fusion options contain a limited number of fusion rules, typically just "and" and "or" rules, so that the search is not extensive. Performance as well as adaptability is restricted due to this small rule set, which either reduces FAR or FRR [12].

In a system targeting the general population, it is paramount that the identification system be able to tailor the collection and matching of biometric data to address the unique characteristics as well as access needs of the individual. This Bayesian framework formalizes the design of a system that can adaptively increase and reduce the security level [13]. This is important to systems designed for varying security needs and user access requirements. The additional biometric modes and the variable error costs give the system adaptability improving acceptability.

6.1 Optimum Bayesian Decision Rule Formation

The problem of personal identification can be formulated as a hypothesis testing problem where the two hypotheses are

H_0: the person is an imposter.

H_1: the person is genuine.

The conditional probability density functions are $p(u_i|H_1)$ and $p(u_i|H_0)$ where ui is the output of the ith biometric sensor given the genuine person and the imposter respectively. The decision made by the sensor i is

$$u_i = \begin{cases} 0, imposter \\ 1, genuine \end{cases}$$

This decision is made based on the following likelihood ratio test

$$\frac{p(u_i|H_1)}{p(u_i|H_0)} \underset{u_i=0}{\overset{u_i=1}{\gtrless}} \lambda_i \qquad (15.1)$$

where λ_i is an appropriate threshold [14-16]. The threshold is assumed to be set internally in the biometric sensor to meet the sensor's design performance criteria.

We define the errors, FAR and FRR, for a sensor as

$$F_{AR_i}(u_i=1|H_0) \text{ and } F_{RR_i}(u_i=0|H_1)$$

The performance of a detector is often represented in terms of receiver operating characteristics (ROC), which is a plot of genuine acceptance rate, GAR, versus FAR. Different values of FAR yield different operating points on the ROC. It should be pointed out that the optimum decision rule could be designed using various performance criteria (e.g. minimum probability of error). In this chapter, we do not consider the design of individual biometric sensor decision rules. We assume that they have been designed and their ROC's are available to us. Based on this information, we study the fusion of the biometric sensor decisions and derive optimum decision fusion rules.

6.2 Biometric Sensors

Three biometric sensors are considered in this chapter: a face recognition system by Visionics, a hand biometric system by Recognition Systems, and a voice recognition system by OTG using the SecurPBX demonstration system [10]. Figure 15-4 contains the three ROCs constructed based from data available in [10]. The biometric sensors selected have comparable performance so that a single sensor does not perform better in both FAR and FRR. As mentioned earlier, human factor issues can affect the performance of a single sensor and are not accounted for in the operating curves.

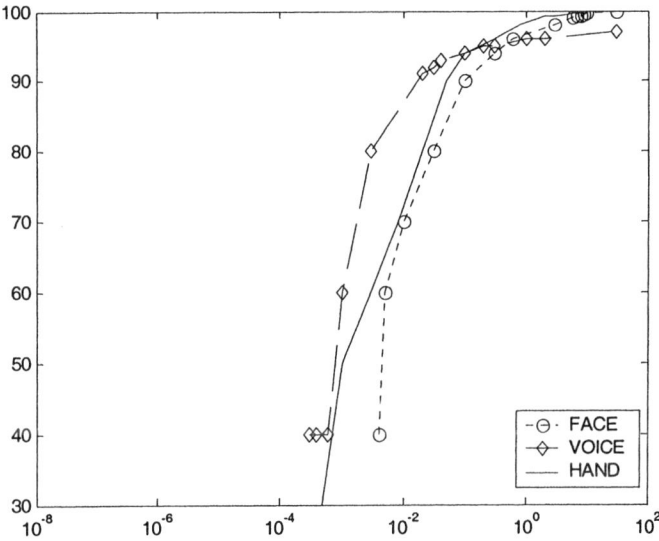

Figure 15-4. Experimentally Determined Operating Curves for 3 Biometrics

6.3 Optimum Decision Level Fusion

Decision level fusion creates a global decision based on the local decisions of the biometric sensors. The system here focuses on three sensors; the results can be generalized to any set of biometric sensors. The system designer's goal is lowering both FAR and FRR by combining the sensor decisions. The error reduction must be significant enough to justify the added cost, complexity and transaction time required by a multiple sensor system.

The optimum decision fusion rule is obtained using the Bayesian framework. Specifically, a weighted sum of the two types of error probabilities is minimized where the weights are the costs associated with the two types of errors. The total cost to be minimized is

$$E = C_{FA}F_{AR} + C_{FR}F_{RR} \tag{15.2}$$

where CFA is the cost of falsely accepting an imposter individual, CFR is the cost of falsely rejecting the genuine individual, FAR is the global FAR, and FRR is the global FRR. This can be rewritten in terms of a cost using

$$C_{FR} = 2 - C_{FA} \tag{15.3}$$

giving

$$E = C_{FA}F_{AR} + (2 - C_{FA})F_{RR} \tag{15.4}$$

The optimum fusion rule minimizes the total cost (4) by selecting the rule that combines single biometric sensor decisions into a combined decision. The single sensor observations and the corresponding decisions are assumed to be independent. Costs have been included in the expression (4) to allow the system designer to either weight the FAR more heavily or FRR more heavily depending upon the application to which he is using the system. Higher cost for false acceptance is typically required for a high security system. High FRR is, however, required for forensic applications in which the system might want to accept everyone who has even a slightest match. Varying these costs affect the fusion rule selection.

The optimum fusion rule allowing access to a building for N sensors is [15][16]

$$\sum_{i=1}^{N} \left[u_i \log\left(\frac{1-F_{RR_i}}{F_{AR_i}}\right) + (1-u_i)\log\left(\frac{F_{RR_i}}{1-F_{AR_i}}\right) \right] \mathop{\gtrless}_{u_g=0}^{u_g=1} \log\left(\frac{C_{FA}}{2-C_{FA}}\right) \tag{15.5}$$

where u_i is the local sensor decision (1 to accept identity or 0 to reject), u_g is the global decision, and N is the number of sensors. There are a total of 2^{2^N} possible fusion rules for the sensors considering all possible combinations of the sensor decisions. The system designer selects a set of individual sensor operating points as well as error costs. Then based on these

15. Optimum fusion rules for multimodal biometric

selections, the optimum fusion rule is obtained. The designer can modify the optimum rule by varying the cost and/or sensor operating point.

First, the fusion of the face and voice biometric sensor decisions is analyzed. There are 16 potential rules possible as depicted in Table 15-2. Most of these fusion rules do not improve performance[16] so that only the monotonic fusion rules need to be considered. The most commonly used rules are *f2* (AND rule) and *f8* (OR rule). The NAND rule or *f9* is worst performing rule and rarely of interest. The *f1* rule simply allows no one in the building and hence can be used during emergencies. Similarly, the *f16* rule allows everyone in which may be used during open house events, etc., but again not for general usage. The system can be switched into single sensor operation by using the *f4* and *f6* rules.

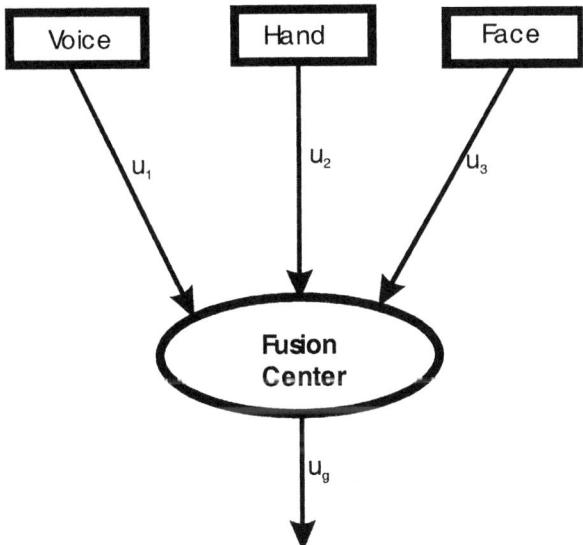

Figure 15-5. Illustration of Fusion of Three Biometrics

In order to gain insight into the performance enhancements resulting from fusion, we present the global FAR and FRR as functions of the individual sensor FAR and FRR for the AND and OR rules. The combined errors for the AND rule are

$$F_{AR} = F_{AR_1} F_{AR_2} \tag{15.6}$$

and

$$F_{RR} = F_{RR_1} + F_{RR_2} - F_{RR_1} F_{RR_2} \tag{15.7}$$

Table 15-2. Fusion Rules for Two Sensors

u_1	u_2	f_1	f_2	f_3	f_4	f_5	f_6	f_7	f_8
0	0	0	0	0	0	0	0	0	0
0	1	0	0	0	0	1	1	1	1
1	0	0	0	1	1	0	0	1	1
1	1	0	1	0	1	0	1	0	1

u_1	u_2	f_9	f_{10}	f_{11}	f_{12}	f_{13}	f_{14}	f_{15}	f_{16}
0	0	1	1	1	1	1	1	1	1
0	1	0	0	0	0	1	1	1	1
1	0	0	0	1	1	0	0	1	1
1	1	0	1	0	1	0	1	0	1

In order to gain insight into the performance enhancements resulting from fusion, we present the global FAR and FRR as functions of the individual sensor FAR and FRR for the AND and OR rules. The combined errors for the AND rule are

$$F_{AR} = F_{AR_1} + F_{AR_2} - F_{AR_1} F_{AR_2} \qquad (15.8)$$

and

$$F_{RR} = F_{RR_1} F_{RR_2} \qquad (15.9)$$

The ROCs for these two rules provide a comparison of the rules' performance. The OR rule ROC curve is steeper with lower FRR for the same high FAR values in Figure 15-6. Thus FAR does not improve but FRR does. Conversely, the AND rule ROC curve is flatter than for the OR rule indicating better FAR performance. An insight gained from this study is that the performance of one of the errors is constrained by the weakest sensor with only 2 biometric sensors. FAR or FRR can be improved through the fusion rule but both can still not be reduced simultaneously.

Thus, a third biometric sensor with comparable accuracy is introduced into the suite. The problem of analyzing all possible rules becomes more complex since there are potentially 256 fusion rules. Once again, however, most of these rules can be ignored since only 20 rules are monotonic [16]. The ROC curves for 4 of these rules are presented in Figure 15-7. These curves clearly illustrate potential improvement in both error types. For added insight, a comparison is made between the ROCs for the AND rule and is given in Figure 15-8 as the number of sensors increases. For improvement in both error types, the ROC must shift from the lower right to the upper left as shown for 1 to 3 sensors.

15. *Optimum fusion rules for multimodal biometric* 279

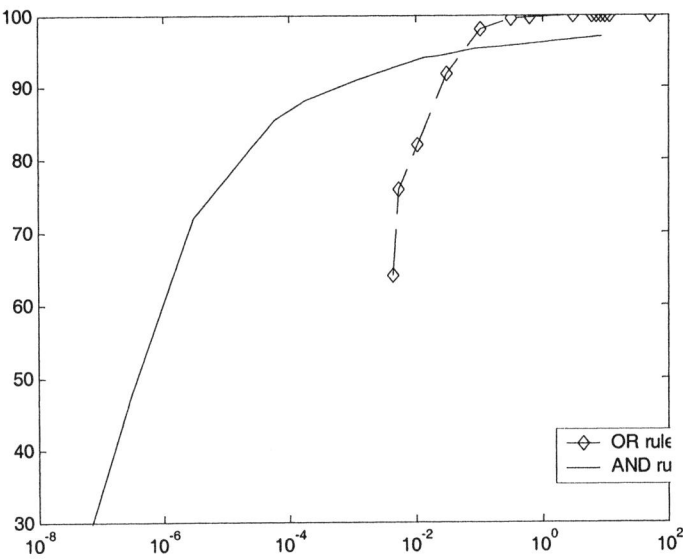

Figure 15-6. ROC for Two Biometric Sensors

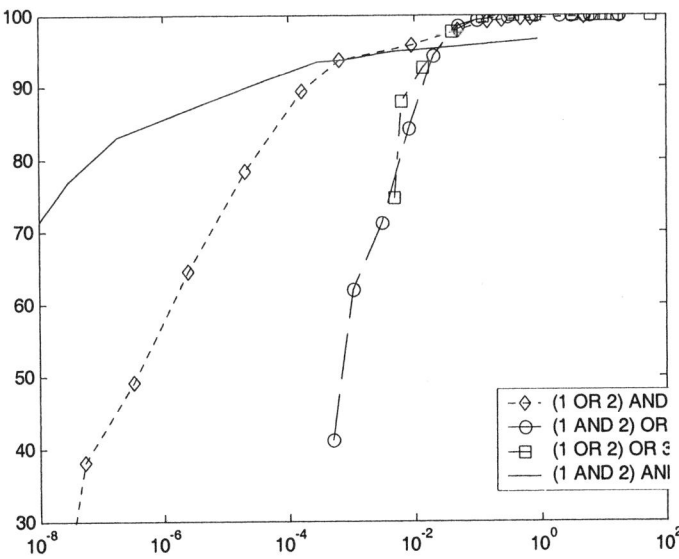

Figure 15-7. ROC Curve for Three Biometric Sensors

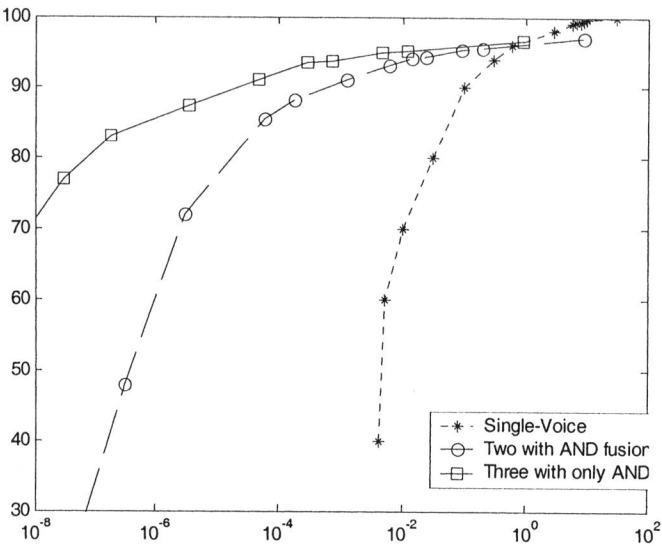

Figure 15-8. ROC Comparison Based on AND Rule

6.4 Impact of Cost on Rule Selection

The system designer increases the impact of a particular type of error by changing the FAR cost in (15-5). The local decisions are first weighted by the accuracy of the sensor's current operating point and then compared with a threshold in (15-6). If the costs are equal, the threshold is 0. The threshold increases as the cost of FAR increases changing the optimal fusion rule to reduce the global FAR. The rule selection process also accounts for the inherent accuracy of the sensor. This results in a table of optimal fusion rules as a function of FAR costs.

Insight into this process is gained by first considering the two sensor problem operating with the same FAR and FRR. It should be mentioned that this is not possible unless the sensors are identical. Logic leads one to conclude that if the FAR is more costly, the AND rule is optimal. If the FRR is more costly, the OR rule is optimal. Table 15-3 shows the optimal fusion rules from an analysis that considered all 6 of the monotonic rules for two sensors. There is a very small range of costs for which allowing everyone and rejecting everyone is optimal. The OR and AND rules transition at a cost equal to 1, i.e., when both errors are equally costly as expected.

A typical set of sensor operating points for the multimodal situation is summarized in Table 15-4. As indicated by the operating points, sensor 1

15. Optimum fusion rules for multimodal biometric

dominates FRR and has a poor FAR. Sensor 2 has better FAR values so the designer should use this to improve performance especially if FAR is more costly. This is accomplished through the AND rule as indicated by (15-7). The table indicates that the AND rule remains optimal for FAR costs down to .7752. Conversely, FRR has to become much more costly with FAR around .6172 before the OR rule becomes optimal. It is interesting to notice the small region of cost values for which sensor 1 should simply be ignored. The optimum fusion rule for FAR costs between .6172 and .7752 relies on only one sensor. Figure 15-9 presents the total error as a function of FAR cost. The lines cross at the points where the optimum rule switches.

Table 15-3. Optimal Sensor Rules As a Function of Cost for Same Operating Point
(FAR1=FAR2=FRR1=FRR2=10%)

Optimal Fusion Rule	Range of Cost Values
All Ones (f16)	0-0.0024
OR (f8)	0.0024-1
AND (f2)	1-1.976
All Zeros (f1)	1.976-2

Table 15-4. Optimal Rule for One Dominant Sensor in FRR
(FAR1=9%,FRR1=0.6%,FAR2=1%,FRR2=4%)

Optimal Fusion Rule	Range of Cost Values
All Ones (f16)	0-0.0005326
OR (f8)	0.0005326-0.6172
Sensor 2 (f6)	0.6172-0.7752
AND (f2)	0.7752-1.9982
All Zeros (f1)	1.9982-2.000

Sensor 1 has a significantly lower FRR than sensor 2 but much higher FAR in Table 15-5. Thus, the OR rule remains optimal even for cost values up to 1.3916. This prevents the FAR to be reduced unnecessarily by the AND rule as shown by comparing (15-7) and (15-9). It is interesting to see that once again there is a range of cost values for which sensor 2 decisions are used. This is a larger range of values. Thus, the operating points of the sensors should be carefully chosen in order to improve system accuracy. Figure 15-10 presents the total error for the fusion rules as a function of FAR cost for Table 15-5.

A selected set of fusion rules is analyzed for three sensors in Table 15-6. This is a much more complicated situation. It is interesting to note that the two middle rules can be used interchangeably for certain cost values. This table does not summarize the entire set of rules but provides insight concerning the advantages of using three sensors. There are also cost regions where it is best to ignore 1 or 2 of the sensors as shown in the previous tables.

Table 15-5. Optimal Rule for One Dominant Sensor in FAR (FAR1=3%,FRR1=0.04%,FAR2=2%,FRR2=7%)

Optimal Fusion Rule	Range of Cost Values
All Ones (f16)	0-0.0028
OR (f8)	0.0028-1.3916
Sensor 2 (f6)	1.3916-1.9592
AND (f2)	1.9592-2.0
All Zeros (f1)	2.0-2

Table 15-6. Optimal Fusion Rules for Three Biometric Sensors (FAR1=6%,FRR1=1%,FAR2=0.1%,FRR2=6%,FAR3=0.8%,FRR3=2%)

Optimal Fusion Rule	Range of Cost Values
Only OR	$2.57*10^{-5} - 0.03917$
(Sensor 1 AND Sensor 2) OR Sensor 3	$2.57*10^{-5}$ -0.14535, 0.3355-1.9836, 1.9934-1.99836
(Sensor 1 AND Sensor 2) AND Sensor 3	0.33557-1.9836, 1.9934-1.99836
Only AND	1.99836-2

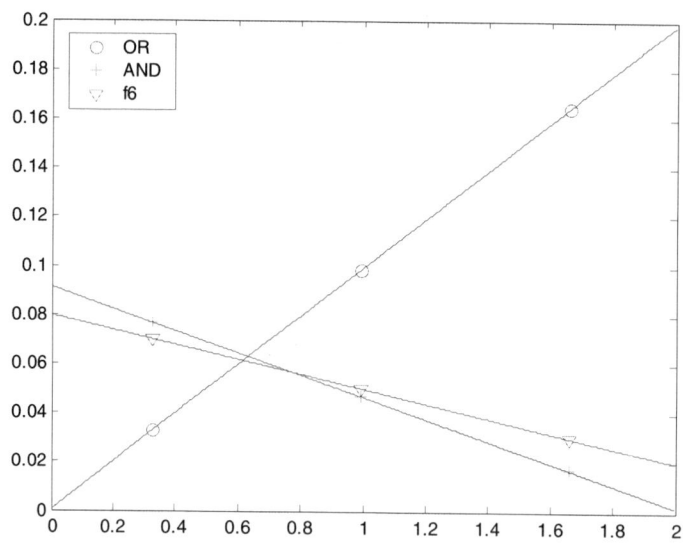

Figure 15-9. Cost Vs. Total Error for the Operating Point in Table 15-4

15. *Optimum fusion rules for multimodal biometric*

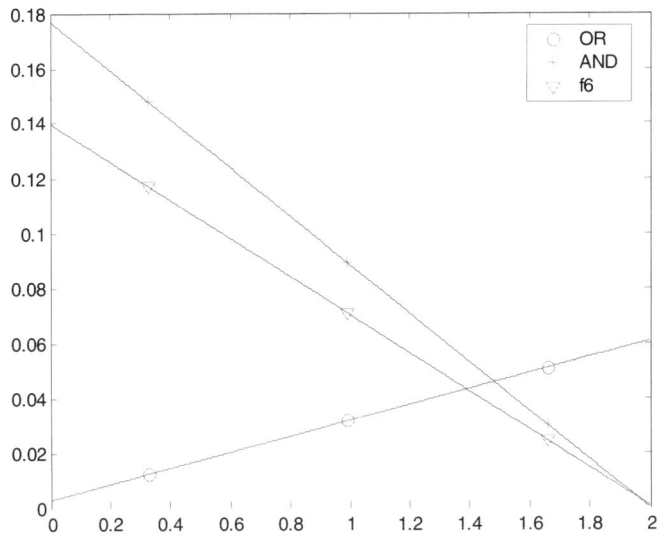

Figure 15-10. Cost Vs. Total Error for the operating point in Table 15-5

7. CONCLUSIONS

It is impossible to address all the human factor and performance issues through a single biometric modality. A multimodal biometric approach using a Bayesian framework gives the system designer flexibility to design a robust system with adaptable security levels to address a variety of building access applications as illustrated in Figure 15-11. This chapter presented a methodology for designing optimal fusion rules based on the biometric sensor decision thresholds or operating points and FAR costs. Also, the biometric sensor operating point must be carefully chosen in order to improve the system accuracy through fusion. If there is a large difference in the biometric sensor's performance, fusion may not enhance the system at all.

An error cost is introduced into the optimization process providing the system designer with a mathematical tool for trading off FRR and FAR. The cost is incorporated at the fusion center. The system operator can change the cost of either error type as the real-time situation changes. This may cause the optimum fusion rule to switch and ultimately change the system performance.

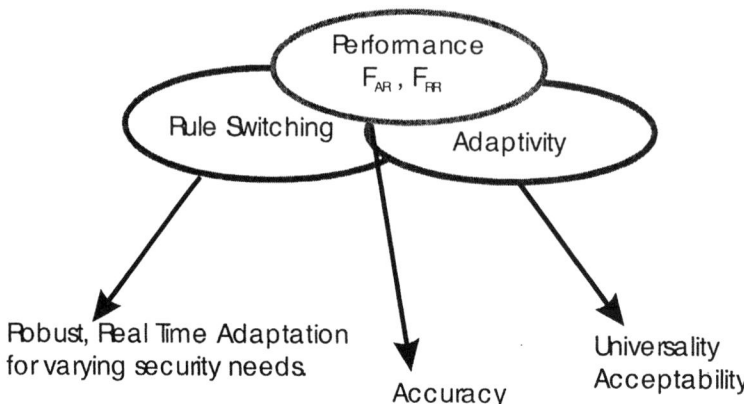

Figure 15-11. Multimodal Biometric Fusion Leads to an Improved System

Thus, a biometric security system based on multimodal biometrics is more robust, adaptable, secure, and user friendly. For users that experience problems with certain biometric acquisitions, the operating points and costs of the individual biometric modalities may be adapted. This allows the system to maintain the same level of security while servicing a population of varying needs. Thus in using a Bayesian framework, multi-modal biometric fusion allows the system to better adapt to the changing security needs of the building as well as address the human factor issues.

REFERENCES

[1] John D. Woodward, Jr., "Biometrics: Facing Up to Terrorism", *The Biometrics Consortium Conference 2002*, Arlington Virginia, February, 2000.

[2] Steven King, "Personal Identification Pilot Study," *The Biometrics Consortium Conference 2002*, Arlington, Virginia, February, 2002.

[3] Lin Hong and Anil Jain, "Integrating Faces and Fingerprints for Personal Identification", *IEEE Transactions on Pattern Analysis and Machine Intelligence*, Vol. 20, No. 12, Dec., 1998, pp. 1295 - 1307.

[4] Salil Prabhakar and Anil Jain, "Decision-level Fusion in Fingerprint Verification", *Pattern Recognition*, vol. 35, 2002, pp. 861-874.

[5] Sharath Panikanti , Ruud M. Bolle , Anil Jain, "Biometrics : The Future of Identification", *IEEE Computer*, Vol. 33, No. 2, February 200.

[6] A.K. Jain, S. Prabhakar, S. Chen, "Combining multiple matchers for a high security fingerprint verifi-cation system", *Pattern Recognition Letters*, vol. 20, no. 11-13, 1999, pp. 1371-1379.

[7] A.K. Jain, R.M. Bolle, S. Pankanti (Eds.), *Biometrics : Personal Identification in a Network Society*, Kluwer Academic Publishers, MA, 1999.

[8] Forrester Research, Inc, http://www.forrester.com, 2001.

[9] Gartner Group, http://www.gartner.com, 2001.

[10] Tony Mansfield, Gavin Kelly, David Chandler, and Jan Kane, Biometric Product Testing Final Report, Computing, National Physical Laboratory, Crown Copyright, UK, March, 2001

[11] Robert W. Frischholz , Ulrich Deickmann, "BioID: A Multimodal Biometric Identification System" , *IEEE Computer*, Vol. 33, No. 2, February 2000.

[12] L.Hong, A.K. Jain, S. Panikanti, "Can multibiometrics improve perfomance?", *Proceedings AutoID'99*, Summit, NJ, October 1999, pp. 59-64.

[13] L.Osadciw, P.K.Varshney, and K. Veeramachaneni, " Improving Personal Identification Accuracy Using Multisensor Fusion for Building Access Control Applications", *Proceedings of the Fifth International Conference on Information Fusion*, July 2002, Annapolis, Maryland.

[14] Steven M. Kay, Fundamentals of Statistical Signal Processing: Detection Theory, Vol. II, Prentice-Hall, Inc., 1998.

[15] Ramanarayanan Viswanathan and Pramod K. Varshney, "Distributed Detection With Multiple Sen-sors: Part I - Fundamentals", *Proceedings of the IEEE*, Vol. 85, No. 1, Jan., 1997, pp. 54 - 63.

[16] Pramod K. Varshney , *Distributed Detection and Data Fusion*, Springer, New York, 1997.

INDEX

acquisition; 31; 44; 48; 49; 58; 105; 124; 235; 253; 267; 268
adaptation; 138; 146; 187; 195
advanced visual-based surveillance; 211
agent; 25; 98; 100; 102; 104
background; 14; 32; 33; 35; 36; 37; 39; 40; 44; 48; 49; 50; 51; 52; 53; 58; 59; 67; 73; 74; 107; 109; 123; 124; 125; 126; 127; 128; 131; 138; 141; 147; 184; 185; 186; 188; 195; 198; 200; 202; 243
background frame differencing; 200
background updating; 202
bounding box; 15; 16; 32; 61; 67; 73; 254
camera calibration; 27; 77; 202
CCD; 31; 51; 137; 201
change detection; 10; 26; 37; 39; 40; 132; 133; 195; 201; 210
classification; 15; 86; 127; 188; 201; 206; 207; 209; 220; 238
CMOS; 268

communication channel; 31; 33; 257
compression; 119; 120; 128; 129; 132; 183; 210
computer vision; 44; 81; 91; 108; 109; 110; 142; 145
display; 16; 105
emergencies; 277
environmental conditions; 110; 235
event detection; 13; 201; 228
events; 8; 13; 34; 39; 61; 62; 97; 143; 183; 201; 203; 205; 208; 209; 211; 219; 229; 277
false alarm; 35; 37; 105; 208; 209
feature extraction; 11; 196; 220; 221; 224; 226; 232; 246; 262; 267; 268; 272
feature space; 242
Gabor; 237
geometric information; 202
ground truth; 20; 21; 22; 24; 72; 73; 76; 78
ground-plane hypothesis; 202
ground-truth; 226

illumination; 11; 30; 40; 68; 120; 122; 126; 132; 185; 187; 188; 190; 195; 251; 252; 260; 261; 262; 269
illumination changes; 126; 185; 190; 251; 252; 260; 261
image acquisition; 269
image processing; 10; 29; 56; 137; 200; 272
infrared sensors; 7
IR; 8; 9; 11; 15; 20; 21; 22; 24; 90; 91; 200
Kalman filter; 14; 17; 27; 72; 73; 184; 185; 202; 225
learning; 62; 66; 67; 77; 187; 188; 199; 200; 201; 202; 205; 206; 208; 263
lighting conditions; 17; 57; 73; 185; 195; 235; 236; 245
localization; 8; 17; 186; 202; 251
Mahalanobis; 56; 237
maintenance; 147
management; 103; 104
Markov; 210; 219; 221; 228; 229; 231; 232
misdetection; 122
missed alarm; 208; 209
motion detection; 53; 58; 74; 104; 105; 107; 123; 125; 183
motion estimation; 200
motion segmentation; 200; 210
moving objects; 12; 15; 25; 30; 31; 33; 35; 40; 108; 109; 120; 122; 123; 124; 125; 126; 127; 128; 132; 185; 200; 203; 210
multi-camera; 26; 48; 77; 136; 143; 145; 146; 183; 184; 185; 189; 190; 191; 192; 197; 219; 221; 232
neural networks; 199; 206

object tracking; 11; 122; 125; 128; 132; 195; 196; 199; 200; 209; 210; 234; 252
operator; 13; 20; 29; 45; 66; 105; 110; 144; 200; 283
outdoor scenes; 201
outdoor surveillance; 31
people counting; 199
performances; 30; 119; 128; 130; 131; 132; 208; 236; 243; 246; 247
Principal Component Analysis; 55; 237; 240
radar; 27
recognition; 8; 12; 15; 25; 26; 32; 34; 35; 54; 135; 142; 190; 196; 200; 201; 202; 203; 204; 205; 206; 207; 208; 209; 210; 211; 220; 233; 235; 236; 237; 238; 239; 240; 242; 243; 246; 247; 248; 249; 251; 265; 267; 268; 269; 275
reconstruction; 135; 210; 222; 223
regions of interest; 120; 123; 132
reliability; 10; 121; 238; 244; 245; 246; 247
remote control; 119; 143
resolution; 9; 20; 24; 29; 31; 32; 105; 127; 190; 191; 195
retrieval; 145; 183; 197; 269
robustness; 261
safety; 200; 266
scenarios; 97; 98; 100; 119; 120; 127; 128; 132; 196; 247
security; 41; 45; 267
segmentation; 11; 17; 21; 108; 186; 197; 220; 268
sensors; 10; 95; 104; 200; 275; 278; 279; 282
signal processing; 200
smart sensor; 10

Index

statistical model; 123; 132; 185; 201; 205; 210
stereo vision; 59
storage; 70; 105; 119; 183; 196
surveillance cameras; 30; 41; 61; 62; 236
system performance; 206; 272; 283
target; 16; 43; 47
task; 41; 44; 50; 52; 53; 57; 188; 200; 201; 203; 205; 206; 208; 226; 245; 246; 247; 269
tracking; 16; 27; 33; 43; 44; 49; 53; 54; 57; 58; 59; 72; 73; 74; 78; 79; 108; 109; 112; 121; 122; 123; 185; 210; 222; 226; 227; 234; 256; 259; 260; 261; 263
traffic; 199; 201; 209
train; 119; 201; 206; 207; 230
trajectory analysis; 12; 13; 20; 43; 199; 205
user interface; 13; 140; 145; 147; 195; 271
video sequence; 21; 53; 57; 59; 93; 120; 121; 135; 145; 184; 193; 194; 196; 197; 210; 221; 228; 230; 236; 238
wavelet transform; 82; 129